Praise for *The Tree of Life*

'Rich with anecdote and infectious enthusiasm, *The Tree of Life* should delight anyone with even a passing interest in the miracle that is life on our planet' Henry Gee

'Combining cutting-edge genetics, a dollop of history and terrifically bizarre creatures, this endlessly entertaining and exciting account is essential reading' Matthew Cobb

'Telford is one of our generation's most brilliant biologists and *The Tree of Life* is a wonderful and vivid guide to evolution's marvels' David George Haskell

'Beautiful . . . a breezy and very accessible way to get readers to think like scientists, and to see the tangled branches of our near and distant relatives all at once' Thomas Halliday

'If you've ever wondered how all of life is related, how we came to be, and how we know, then this brilliant and beautifully written book is for you. The greatest story ever told, presented with exemplary clarity and style' Tim Blackburn

'An acrobatic exploration of the mightiest branches of the tree of life. Bursting with history and natural history, and dangling with tales of clever sleuthing, this is an authoritative and personal guide to life's evolution and our own origins' Nick Lane

'A fabulous tour de force on the evolution of the natural world; through cells, fish, humans, worms and more, Telford reveals the complexities of life with an extraordinary mix of humour, history, wit and erudition' Seirian Sumner

The Tree of Life

The Tree of Life

Solving Science's Greatest Puzzle

MAX TELFORD

JOHN MURRAY

First published in Great Britain in 2025 by John Murray (Publishers)

1

A CIP catalogue record for this title is available from the British Library

Hardback ISBN 978-1-399-80637-4
Trade Paperback ISBN 978-1-399-80638-1
ebook ISBN 978-1-399-80640-4

Typeset in Bembo by Palimpsest Book Production Ltd, Falkirk, Stirlingshire

Printed and bound in Great Britain by Clays Ltd, Elcograf S.p.A.

John Murray policy is to use papers that are natural, renewable and recyclable products and made from wood grown in sustainable forests. The logging and manufacturing processes are expected to conform to the environmental regulations of the country of origin.

Carmelite House
50 Victoria Embankment
London EC4Y 0DZ

www.johnmurraypress.co.uk

John Murray Press, part of Hodder & Stoughton Limited
An Hachette UK company

The authorised representative in the EEA is Hachette Ireland, 8 Castlecourt Centre, Dublin 15, D15 XTP3, Ireland (email: info@hbgi.ie)

For Lorna, Celia, Seth and Francesca

Contents

Introduction

It's october, and the view from my desk gives onto the seasonally misty farmland of south-west Dorset. I can see a small grassy garden surrounded by shrubs, brambles, roses and apple trees. The lawn has moles living beneath it, and in the summer, I saw a hedgehog, a grass snake, birds (sparrows, chaffinches, tits, buzzards, chiffchaffs . . .), butterflies, beetles, wasps, snails and earthworms. Beyond the garden are farmers' fields, where grasses, sunflowers, rape and linseed are growing, and woods with foxes, pheasants, trees, edible fungi, poisonous fungi and countless kinds of arthropods. In the distance, the hillside is filled with Jurassic ammonites, belemnites, ichthyosaurs and crinoids. If I could peer through the hill, I'd see the waterfowl on the Fleet Lagoon and then, beyond Chesil beach, the English Channel where there are seaweeds, dolphins, dogfish, mackerel, cuttlefish, lugworms, roundworms, flatworms, penis worms, fat-innkeeper worms, spoon worms, arrow worms, I could go on . . .

A little careful looking or listening or even sniffing in almost any environment on earth will tell you that life is dazzlingly diverse. And the glimpse I have from my window ignores thousands of shyer, rarer and smaller species of animals, plants and fungi and millions of invisible, single-celled creatures hidden in the fields, woods and waters around me, from algae to bacteria. It's not just the number of species that is breathtaking but also the variation between them; even the impossibly partial picture of the species that live within a mile or two of my desk reveals countless unique characteristics – feathers, bones, shells, seeds, chloroplasts, xylem and phloem, muscles, brains, claws, teeth, horns and a dozen different kinds of eye. In fact, the list of inventions is even more awe-inspiring than the long roll-call of species that possess them.

If we can follow its many threads, the story telling *how* all of this diversity arose – both the many species and the all but uncountable characters they possess – will be the history of the most extraordinary series of events in the universe.

You probably know the outlines of our own story: the ancient and humble origins of life; the evolution of more complex cells; the first animals; fish then amphibians; mammals then apes; neanderthals then *Homo sapiens*. But have you wondered how we came to know even this brief account of our own history? Can we discover when muscles and teeth and fur evolved? And know what else was happening on the planet when they did? And what about the biographies of the other billion or so species of life?

Darwin's and Wallace's theory of the origin of species by natural selection provides the mechanism for how evolution works – the cogs and wheels of the process that gave giraffes their long necks and stole our tails. While obviously (obviously!) hugely important, this doesn't tell us the true history behind the appearance of the life outside my window. My focus here (with Darwin's mechanism whirring away behind the scenes) is on how we might come to tell the rollicking history of life. This is an alternative account of evolution that aims to tell us about the events that *actually happened*, where they happened, when they happened, the players involved, who said what to whom. We'd like to know about the accidents and coincidences that produced today's biodiversity – the unforeseen consequences of the evolution of teeth or the effects of volcanoes, meteorites or viruses on the diversity of life today.

With such an account of the story of life on earth as our ultimate goal, this book will explain just how these events can be known at all. I will show how telling the history of life depends on the epic task of reconstructing the tree of life – a family tree that captures how all species of life from oak trees to orcas are connected. This tree of life is a visual representation of relatedness whose intuitive simplicity belies its vast descriptive power.

The first part of the puzzle is to discover how the many different species that have appeared over time are related to each other. At some point, soon after the first appearance of life, a single ancient species, the ancestor of everything that came after, divided to make two separate species. As the tree grew and time passed, these two pioneers would go on to divide again and again, producing more and more species (and more and more branches of the tree). Reconstructing this huge family tree is the first of our tasks, but a family tree of anonymous ancestors is rather dull. We want to ask of the tree of life the same questions we ask of a family history: who was a king and who a convict; what did they look like; what was their character; when and where and how did they live? A tree on its own is like a perfectly pieced-together jigsaw puzzle with no pictures on it.

Our next task, then, is to colour in these blanks, to decorate our huge genealogy with the rich and elaborate details of the biology of the many members of the family of life: we want to know about the evolution of their genes, their morphology and behaviour, the lucky breaks, the catastrophes dodged (or not) and the influence of all the other species – predators, prey and parasites.

As time passed, every branch within the tree evolved new characteristics that, accumulating generation after generation, have produced the present diversity of living beings, a tiny sample of which I glimpse from my study window. Some of these characteristics are striking, like the backbone found in all vertebrates, or the flower unique to angiosperms. Many others are much more subtle: in fruit flies, the pattern of pigment on the wing, minute chemical variations of odour, and the precise frequency of wing beat of the male's mating dance can all be discerned and act to distinguish one species from another. To include this second part of the story is to add flesh to the bare skeleton of the tree's form.

When we combine these two elements – the pattern of species' relationships and the evolution of their characteristics – we come to understand that each species has its own genealogical history of closer and more distant relationships to all other species, and that each, as we trace its path through time and up the tree of life, has accumulated its own unique library of characteristics. I got my backbone on the branch leading to vertebrates, gained nipples on the branch leading to the mammals and later lost my tail on the branch leading to the apes.

I want to reveal how a worldwide community of scientists is cooperating to build the tree of life to tell the complete history of the evolution of life's diversity. We will discover that it is the properties of living species themselves that are the clues we use to discover the true shape of the tree of life. We will see how these characteristics, the products of the process of evolution, are both the raw materials for tree building *and* the very objects whose evolution we wish to explain.

These concepts will lead us to understand how we are able to use the tree of life to go back in time, extrapolating backwards from living species to reconstruct long dead ancestors. Along the way we will find that evolution is often unpredictable and that the unexpected byways – the exceptions to the rules – are the source of the mistakes we sometimes make when building the tree of life. These anomalies, frustrating though they can be, are also where some of the most interesting and surprising passages of the history of evolution are written.

I am a zoologist and I spend my working life trying to reconstruct the tree of life (the part that encompasses the animal kingdom at least). In the following pages I will try to explain why solving this puzzle is so important. What we are reaching for with the tree of life is nothing less than a complete, multi-billion-year history of the emergence of all life on earth in all its extraordinary complexity. My ultimate motivation comes from knowing that the tree of life is a portal that can transport us back in time to meet our ancestors. Once we have pieced together the clues to reveal a tree that we trust, we can transport ourselves back in time to land at its root. From here we can clamber up to follow the sequence of improbable events that led through 4 billion years of evolution from the first simple cell to an ape that can wonder about his origins.

PART I

What Is the Tree of Life?

1

Solving Science's Greatest Puzzle

In the spring of 2022, Cambridge University reported the mysterious reappearance of a pair of priceless artefacts – two of Charles Darwin's notebooks ('B' and 'C'), jotted and sketched in by Darwin in 1837, early in the long gestation of his theory of the origin of species. The notebooks had disappeared twenty-two years previously, when the university had digitised its Darwin archive – their loss not immediately noted due, perhaps, to the chaos of the digitisation project. The notebooks had been stored together in a small paperback-sized box and were at first presumed lost somewhere on the 130 miles of shelving in the vaults of the university library, but repeated searches over almost two decades turned up nothing. The university librarians admitted defeat in 2017 and, slightly red-faced and sweaty-palmed, one imagines, finally reached out to the public for information. 'We are appealing for anyone with any knowledge of the whereabouts of these priceless artefacts to contact us. They are extremely valuable and important, both to the university, and anyone interested in the history of science,' said Detective Sergeant Sharon Burrell of Cambridgeshire Police.[1] DS Burrell was not exaggerating the importance of these notebooks. In notebook B lies the earliest known diagram of an evolutionary tree drawn by Darwin – what we now call the 'Tree of Life' sketch. The notebooks, still in their container and in perfect condition, reappeared on 9 March 2022, left in a gaudy pink bag for the attention of the librarian; the identity of the thief remains unknown.

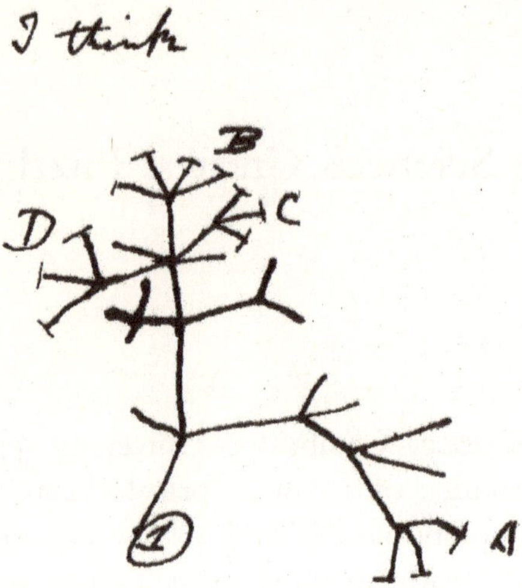

FIGURE 1: Darwin's 'Tree of Life' sketch from notebook B (1837).

Darwin's famous tree of life (annotated with the gnomic comment 'I think') is small and simple but manages, nevertheless, to be an accurate prototype of the trees we use today. Its simplicity is precisely what makes it such a perfect diagram, easily understood but capable at the same time of conveying a huge amount of information. At the bottom we can discern a root that tells us about the long-dead ancestor that would spawn all the species on the tree. Upwards from this root grows a trunk whose extent shows us the passing of time. At some point the trunk divides – a single species becoming two and these two evolving to become different. Each of these sister branches goes on to divide again to produce more branches. Finally, at the very tips of each branch, we find species. The branches of the tree show us, most importantly of all, the relationships between these species. Sibling relatives emerge from the most recent splits in the tree, more distant relatives from splits deeper in the tree. To encompass all of life and the entire history of evolution, we need do nothing to this tiny tree beyond scaling it up to add more species related by more branches. The rules of its construction and interpretation and the information it conveys remain perfectly intact as it grows.

Trees of life are much older than Darwin's first effort. Their earliest uses were as symbols in art, literature and especially religion across many cultures – Norse, Mesopotamian, Chinese, Zoroastrian – and their meanings, often obscure, are as diverse as the peoples who described, painted and carved them.[2]

The meaning of Darwin's tree of life is utterly different from these mystical symbols but, even in the absence of mystery or metaphor, what it tells us is extraordinary. The word 'life' refers not to a human lifespan (a life), nor simply to the state of being alive (life as opposed to death); rather it is an almighty collective noun encompassing every single living thing, every species alive today and every species that has ever existed. The tree that links this amazing collection of beings is a rich diagram, a perfect representation of the relationships – the *evolutionary* relationships – between every one of them.

Allowing for nineteenth-century language, Darwin, of course, gives a beautifully clear explanation of the use of a tree to represent evolution:

> The affinities of [relationships between] all the beings of the same class have sometimes been represented by a great tree. I believe this simile largely speaks the truth. The green and budding twigs may represent existing species; and those produced during each former year may represent the long succession of extinct species.

I confess I am quoting selectively, because the full version of Darwin's tree of life strays much closer to a simile for the *process* of evolution than the modern tree of life does. Darwin also uses his tree to represent the competition between species – the struggle for existence and the survival of the fittest:

> At each period of growth all the growing twigs have tried to branch out on all sides, and to overtop and kill the surrounding twigs and branches, in the same manner as species and groups of species have tried to overmaster other species in the great battle for life.[3]*

* Surprisingly, while Darwin had a long-standing interest in using trees to illustrate the process of evolution (and was jotting trees in his notebooks decades before

The modern tree of life abandons any attempt at metaphor or simile for the process of evolution and seeks more simply to represent the history of evolution. While we may have lost a bit of colour, we have gained considerably in the clarity of the tree's message. The simple goal – of representing relationships – is certainly a worthy one, but the knowledge of relationships that the tree contains proves to be much more important than this. The tree can be thought of as the plain steel framework of a skyscraper; it is something into which we can insert the walls and floors and windows of the building – marble and glass and tiles and gargoyles – to produce a much more elaborate edifice. For the tree of life, this elaboration includes the changing characteristics of species, of their ancestors, of births, deaths, invasions, extinctions, dates, geological context, mergers and acquisitions.

Trees showing relationships between species took on a special significance following the publication of *On the Origin of Species* in 1859, but trees representing relationships were already familiar to the Victorians. Genealogical trees – family trees – are a part of human cultures around the world. They can be found in the 'stemmata' (garlands) of ancient Rome; in the eleventh-century 'Confucius genealogy' (still being added to after eighty generations); and in the Tree of Jesse of the Old Testament. The ubiquity of genealogical trees across cultures tells us that they are an inevitable product of human imagination and must have existed deep into prehistory. Genealogies have even been explicitly thought of as tree-like for millennia: 'And there shall come forth a rod out of the stem of Jesse, and a Branch shall grow out of his roots' (Isaiah 11:1) – this biblical family tree leads from Jesse, through (according to Luke) forty-three generations (all men of course) to Jesus, and Christian illustrations of the Tree of Jesse are drawn as literal trees. The genealogical tree of the Cancellieri family of Pistoia (Albero Genealogico dei Cancellieri di Pistoia) engraved in 1581 by the industrious and well-paid sycophant, Scipione Ammirato, shows a

publishing his theory), *On the Origin of Species* contains just one diagram of a tree and it is a fairly plain one at that, its leaves, ignoring the obvious possibilities of aardvarks and zebras, asters and zinnias, are the symbolic species 'A to Z'.

mighty oak riven in two at its trunk, the two boughs representing the violent schism (around 1300) between the white and black Guelph factions of the family. This tree is naturalistic, growing within a Tuscan countryside, two armies bearing black and white standards visible to either side of the tree.

Genealogical trees are familiar and their interpretation simple: the leaves at the tips of a family tree represent individuals of the most recent generation – brothers and sisters and their cousins. The siblings in a family are connected to each other by their shared parent, a part of the bough immediately below them in the tree and just one generation further back in time. A child will connect to its cousins (more distant relatives of course) via an older generation that lies deeper in the tree: the closest ancestor in common between first cousins is not a parent but a grandparent. In this simple way, the degree of relationship between all the leaves of each generation (siblings, cousins, second cousins, parents, uncles and aunts and so on) can be read in the arrangement of the branches that connect them.

Tree diagrams have survived for millennia because they have proved an ideal way to represent this information about relatedness of family members, but tree-like diagrams have been adopted in many other contexts precisely because they are such a natural way to organise *any* set of objects – rocks, stamps, jazz musicians – that can be grouped (or classified) according to how similar or how closely related they are. For the natural world – for species – the desire to bring order to their study meant that, long before it occurred to Darwin that species might be related to each other like members of a family, there was a desire to classify and to organise species.

'The urge to classify is a fundamental human instinct; like the predisposition to sin, it accompanies us into the world at birth and stays with us to the end,' wrote Tindell Hopwood (hyperbolically) in 1959.[4] It is true that we all have an innate skill that allows us to make useful generalisations about the species we encounter; it takes no effort or specialised training for any of us to know that, despite their obvious differences, eagles and pigeons are birds, tigers and sheep are mammals and oak trees and blackberry bushes are plants. Beyond this it is also obvious that mammals and birds can be sensibly lumped together as animals, which is a group to itself that excludes plants. As we intuit all this, our brain is constructing an accurate 'folk' classification of these six species into groups of similar organisms. Notice how this classification (birds, mammals, plants) can be translated directly into a tree – picture a bird branch (that forks to the eagle and pigeon) and a mammal branch (tiger and sheep); connect these two lower in the tree to form a larger animal branch; finally the animal branch connects, down on the trunk of this little tree, to a plant branch that leads to oak and blackberry. The *classification* of life that comes so easily to us (mammals, birds, animals, plants) existed long before any *tree of life*, but the classification and tree are one and the same – it just took a while before it occurred to anyone to illustrate a classification in the form of a tree.

Prefiguring the first trees of life by more than two millennia, the earliest recorded *classification* of life, of animals at least, is credited to Aristotle. His great works *Inquiries on Animals* (usually known via the Latin, *Historia Animalium*) and *Parts of Animals* (*De Partibus Animalium*), written in the fourth century BC, contain his deliberate and characteristically thorough attempt to know and to understand animal life through its many different characteristics. There is a danger of reading Aristotle anachronistically, of attributing to his work a kind of proto-evolutionary viewpoint. He was emphatically not an evolutionist, but these books nevertheless laid the groundwork for modern classification, bringing together both the necessary data and a handful of essential ideas. His data take the form of fantastically detailed and generally accurate observations of the characteristics of a great number of animals (and not just the cuddly, fierce, weird or otherwise charismatic). His interpretations come from a simple philosophical interest in classifying things. The point of Aristotle's classification was to know what each animal *was* through the collection of its attributes and the things it can do, 'to reach the definitions of the ultimate form'.[5]

Hopefully avoiding the trap of reading Darwin between the lines of Aristotle, there are still useful lessons to be found, perhaps the most important of which is that it might be possible to classify animals into groups which are both useful (allowing us to make generalisations) and which we can all agree on – a truth to be discovered. Aristotle's biggest and boldest choice in *Historia Animalium* is to split all animals into two *megista genê* ('big groups'): those with red blood (*enhaima*), corresponding to vertebrates; and those without (*anhaima*), corresponding to invertebrates. Within each of these big groups he recognises a series of lesser groups: within the red-blooded *enhaima* he distinguishes viviparous (live-bearing) and oviparous (egg-laying) quadrupeds (respectively mammals and reptiles/amphibians) as well as groups of whales, fishes and birds. In the bloodless *anhaima* we can discover groups including the *malakia* (soft body – molluscs), *malakostraca* (soft shell – crustaceans) and *ostrakoderma* (a hard shell surrounding the body), which is an unholy mix of sea urchins, sea squirts, shelled molluscs and hard-shelled arthropods like barnacles.

Modern zoologists are naturally delighted to find a series of 'kinds' that we would accept as legitimate groups today; the agreement between ancient and modern may seem unsurprising, but it certainly didn't have to be the case. Rather than using feathers to define a group of birds and milk to group all mammals, Aristotle could have chosen to make a grouping of the terrestrial animals (uniting pigeons, lizards and humans) and another of aquatic animals (ducks, salamanders and fish). As we will find out, the pleasing correspondence between Aristotelian and modern groups of animals was not a fluke but the inevitable consequence of how evolution works.

Aristotle's systematic approach to the study of nature was slowly forgotten by Western scholars, until he was rediscovered in the twelfth and thirteenth centuries. In the following half a millennium, most of the revived interest in classifying life focussed on botany for the very practical reason that accurately identifying plants was essential for their use in medicine. The eighteenth-century Swede, Carl (or Carolus or occasionally Charles) Linnaeus (or Linnæus or von Linné or a Linné) was one such botanist, as well as being by far the most important pre-evolutionary classifier of life. Linnaeus is rightly known as the father of modern classification. His first great innovation was a broad, evidence-based classification of plants and animals into smaller and smaller and more and more exclusive groups. Linnaeus's magnum opus *Systema Naturae* is much more than a classification; it also contains detailed descriptions of the characteristics of individual species and of the groups to which they belong. The overall result was a key that allows a researcher to classify any species to its own increasingly exclusive kingdom, phylum, class, order, family, genus and species, homing in on its unique place in the classification (and ideally thus avoiding poisoning the herbalist's patient). Linnaeus's second great innovation was a formal system for naming species whereby each is given a unique two-part name – a so-called binomen such as *Homo sapiens* – without which modern biology would be utterly chaotic.

The Linnaean system of classification survives largely intact today. It can be compared to the address of your house. Naming just a few, nested geographic groups – the continent, the country, the county or region, the town, the district, the street and the house number – can succinctly pinpoint any house in the world. The Linnaean equivalents of these levels of organisation begin with the kingdom (the animal kingdom, the plant kingdom, etc.), down through phyla (within animals there are the chordates, arthropods and molluscs, for example), then classes (within the chordates are mammals, reptiles and amphibians) and then orders (within mammals are primates, bats and rodents). Orders contain families (primates contain lemurs, monkeys and apes), which contain genera (apes contain *Homo* and *Gorilla*), which each contain one or more species (*Homo sapiens*, as well as the extinct *Homo erectus* and *Homo neanderthalensis*). Your own Linnaean classification is therefore kingdom Animalia (though we now use 'Metazoa'), phylum Chordata (approximately animals with a backbone), class Mammalia, order Primates, family Hominidae, genus *Homo* and species *sapiens*.

While Linnaeus had invented a brilliant system of classification (which we still use today), the evolutionary tree of life was still out of reach – classification and genealogical diagram were yet to meet. This said, the inherent tree-ness of some early classifications becomes, at times, achingly close to an actual tree, the intellectual leap to representing a classification in a tree-like image not quite made. The divisions and their subdivisions are often represented in a table form. In the first column are rows of a few large groups (which would be the big boughs at the base of a tree), each of which is subdivided into smaller groups in the next column (which would be the smaller branches).

Eventually, and long before Darwin's doodles, these tables of classification finally morph into diagrams with the form of a tree, although not before some musing on the possibility that this *might* be done. Swiss naturalist Charles Bonnet, in his 1764 book *Contemplation de la Nature* (I don't think I need to translate that), asked: 'Does the scale of nature become branched as it arises? Are the insects and molluscs two parallel and lateral branches of this great trunk?'[6]

In 1766, Georges-Louis Leclerc, Comte de Buffon, the great French naturalist, likewise thought of the mammals as appearing 'to form families in which one ordinarily notices a principal and common trunk from which seem to have issued different stems and the more numerous, as the individuals of each species are smaller and more prolific.'[7] Finally, in 1801 we find the first explicitly tree-like diagram to record the classification of species. It was drawn by the French naturalist Augustin Augier, who says of his diagram:

> A figure like a genealogical tree appears to be the most proper to grasp the order and gradation of the series or branches which form classes or families. This figure, which I call a botanical tree, shows the agreements which the different series of plants maintain amongst each other . . . just as a genealogical tree shows the order in which different branches of the same family came from the stem to which they owe their origin.[8]

Darwin's theory of evolution by natural selection upended the way we think about both classification and its representation in tree form because it implied the existence of a single, perfect, historically accurate classification. Yet *On the Origin of Species* was mainly intended as an explanation of the mechanism by which new and different species appear; it contains just one diagram of a tree, and it is a fairly plain one at that, merely demonstrating how species divide and go extinct. Nevertheless, Darwin understood the tree of life's taxonomic potential, as he expressed in a letter to T. H. Huxley in 1857: 'The time will come I believe, though I shall not live to see it, when we shall have very fairly true genealogical trees of each great kingdom of nature.'[9]

While Darwin did draw a handful of evolutionary trees, he seems to have preferred the idea of representing the evolutionary history of life as a coral, with the dead, stony interior representing the past (and the fossil record) and only the living tips of the coral representing the species that are alive today. Darwin's German disciple Ernst Haeckel, on the other hand, was a big fan of a tree of life. Haeckel was born in Potsdam near Berlin in 1834 to professional parents, Carl (a high-ranking local official) and Charlotte (née Sethe, daughter of a privy councillor). Haeckel's beloved and short-lived first wife, Anna (also née Sethe), was a cousin on his mother's side of the family. Like Darwin, Haeckel, at the insistence of his father, trained initially as a medical doctor but abandoned medicine soon after meeting his first patients.[10] Perhaps the research for his doctorate in medicine, which was awarded for a study of crayfish, left him ill-prepared for the realities of the human body, rather like the story of John Ruskin on his wedding night.

A pair of famous early photos show Haeckel in 1866, prior to a collecting trip to the Canary Islands, with his assistant, the Russian Nikolai Miklouho-Maclay. Even though the photographs were taken not in the field but in the German university town of Jena, the two explorers have dressed in expedition clothes and are surrounded by the paraphernalia of professional collectors – butterfly nets, a bucket, other bits of equipment whose uses I cannot guess at, a dried starfish and a crab on the floor and a dead brittle star in Haeckel's hand. The photographs are carefully posed, and the two men really seem to relish the idea of themselves as tough explorers. In the first, Haeckel, his youthful beard rather wispy, is slouched on a chair, while Miklouho-Maclay stands with a raincoat hanging casually from one shoulder, his left hand resting on what looks like a dagger, while he smoulders at the camera. In the second, they have swapped places and Haeckel, now standing, has taken off his shoes and socks and rolled up his trousers while Miklouho-Maclay has messed up his hair. Photographs from later years include one of a now middle-aged Haeckel surrounded by jungle plants and dressed for the tropics, pith helmet and all, but again taken in a studio in Jena. And finally, Haeckel in old age is standing Hamlet-like with a human skull in his hand and an ape skeleton looming behind him. Big beards may well have been commonplace at the time, but the resemblance of Haeckel's magnificent beard to that of his hero Darwin in this late photograph is very striking.

Haeckel has had an outsized impact on biology and on the study of evolution. He coined terms for huge new concepts such as ecology and phylogeny (the study of evolutionary relationships through time – hence 'phylogenetic tree'); he wrote a bestselling popular science book explaining his view of evolution; he notoriously tweaked his figures of animal embryos to suit his theories of how the stages of embryonic development ('ontogeny' – another Haeckelian coinage) reflect an animal's evolutionary history ('ontogeny recapitulates phylogeny', in his words); he described never-to-be-seen-again species that conveniently supported his theories; and he painted exquisite and rightly famous pictures of creatures great and small, which are said to have influenced the Art Nouveau movement.

His artistic skills are clear in his most famous evolutionary trees, which, unlike almost all more modern trees you might see, are drawn as real trees with twisted branches and knotted bark. The first such *stammbaume* (family trees or pedigrees) appear in his 1866 treatise that has the not very snappy title of 'General morphology of organisms: general outlines of the science of organic forms, mechanically grounded on the theory of descent reformed by Charles Darwin' (which is usually simply referred to as 'Generelle Morphologie').[11] While pre-Darwinian evolutionary trees exist (they are evolutionary in the sense that they allow for species to change through time), Haeckel's trees have the distinction of being the first drawn with the Darwinian mechanism of evolution in mind, and with the fundamental understanding that all of life could have a single origin.

The first of the series of trees in 'Generelle Morphologie' encapsulates the whole of life as then understood, fitted into a tree with three great branches: plants (in which he includes fungi); protists (mostly comprising tiny, single-celled organisms including bacteria, although he also includes large, many-celled sponges); and animals. These three branches are shown joined to one another close to the base of the tree, and the whole undivided trunk emerges from a single, universal root of organisms: *Radix communis organismorum*. Trees in subsequent figures deal with the relationships of species within each of these three great boughs – a tree for plants, a tree for animals, etc., and these are followed by zoom-ins to the details of still smaller branches, such as a tree of the mammals. All of Haeckel's tree-like trees in 'Generelle Morphologie' seem to have been drawn to resemble a real species – the 'German oak', which was supposedly chosen because the oak is the highest-ranked plant on the mediaeval *scala naturae*, found immediately below the simplest animals.

At this point, I have to resist the temptation to unpick Haeckel's successes and failures in the accuracy of the relationships he shows (especially on the animal branch) and consider instead what his tree is able to tell us about evolution. To boil it down, there are four features – either implicit or explicit – that we can read on Haeckel's tree: first, the tree shows the passage of time, at least to the extent that parts of any given branch that are closer to the root of the tree represent species that lived longer ago than those represented by parts of that branch that are closer to the top; second, the trunks, boughs, branches and twigs all represent continuous lineages of species that existed in the past; third, there are points where a branch divides in two, where a single ancestral species split into two descendent species; and finally, only the very tips of the tree – or perhaps the leaves – represent species that are alive today.

Unlike the biblical account of the simultaneous creation of all individual species (which, if it used the same representation, would resemble a field of grass), Haeckel's tree of life is a single plant with a single root. It says that there was a species that lived long ago that is the ancestor of all the species that followed. As we move up the tree, any slice through any branch would represent a snapshot of all the individuals of a species alive at that point in time. A fraction further up the branch would represent their offspring, the next generation.

Eventually a branch may divide into two; each branching point records the brief period of time (it was unlikely to be instantaneous) when an entire population that made up a single ancestral species split into two species. Importantly, a species must first divide into two isolated populations (perhaps separated by a river or mountain range) before these can then become separate species. From the point when the two populations become isolated, they are free to evolve independently of each other and in different directions (their genes are no longer being continually mixed up by sex). The result of this ability to change independently is two distinct species, and we call this creation of two species from one 'speciation'.

The more time that passes, the more different the species are likely to become. What this means for Haeckel's tree is that species whose branches separated very low down in the tree and thus very long ago have had a very long time to become very different from each other. The earliest (lowest) divisions on Haeckel's tree separate plants from animals, while those branches that split close to the top of the tree separate much more similar species – for instance chimpanzees and humans.

In a decision that would be frowned on today, Haeckel's tree quite deliberately treats certain groups as special. At the tops of his three great branches, he placed what he obviously viewed as the most advanced members of each. For the animal bough, the topmost branches of the tree are not the molluscs or the earthworms but the birds and mammals, and the topmost branches on the plant bough are the flowering plants (angiosperms), with algae, lichens and fungi relegated to an inferior place. Haeckel can only achieve this effect by taking liberties with the relationship between the passage of time and the length of branches. Because the root represents the single common ancestor from which all modern species have evolved, and the tips of the branches represent species that are alive today, all living species have equally long histories and so should really be placed equally distant from the root.

Trees of life quickly became less beautiful but more practical than Haeckel's; over the following century and a half we have developed conventions for how trees of life should be drawn that mean we can reliably interpret them. To show some of what a tree of life can tell us, figure 2 is a simple example that involves four familiar animals: a cow (a placental mammal), a duck-billed platypus (a charming but rather bizarre egg-laying mammal or *monotreme*), a goldfish (a bony fish) and a butterfly (an insect).

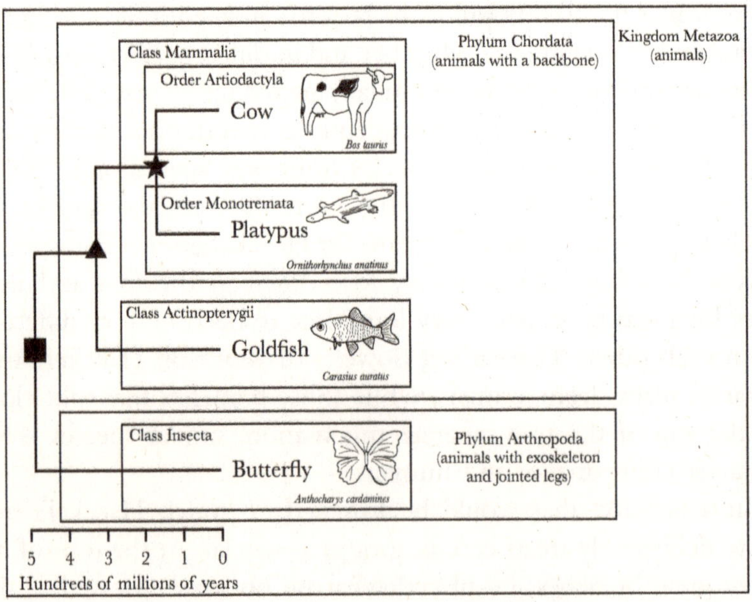

FIGURE 2: What we can read from a simple tree relating four animals.

Fortunately, trees being an excellent way of showing evolutionary relationships, it is easy enough to understand the most important thing that this tree shows, which is the relationships between these four animals. First, the cow and the platypus are each other's closest relatives. The next closest species to both cow and platypus is the goldfish, to which both mammals are equally closely related. Finally, the butterfly is the most distant from the other three species, and again, it is equally distantly related to the three others. We can check this by measuring the total horizontal distance to each species from the base of the tree; all these horizontal distances, which show how much time has passed, are identical. (The lengths of the vertical parts of the tree, in contrast, tell us nothing and are drawn arbitrarily to make the figure easy to read.) While my tree, unlike Haeckel's, obeys the rule that branch lengths indicate the passage of time, I haven't attempted to make the relative distances along different branches correspond to the amount of time that has actually passed. The ancestor of the two mammals was not half the age of the ancestor of mammals and fish, as the lengths of the branches on my tree imply.

We can also read the tree from the bottom up (though the bottom of the tree is located on the left in this figure), starting at the root indicated by the square, to follow the events of evolution that resulted in the four species alive today. The square represents the common ancestor of all the four animals shown on the tree – an ancestor that we can only imagine since it has left us with no trace of its existence. It is equally distantly related to all four living species. What this means is that, even though we might think of the insect as being the most simple, it is in truth no closer to the animal ancestor than the complex mammals are.

The single, unknown species that was the animal common ancestor at some point in time divided into two new species which, once independent of each other, began to diverge – to evolve to become increasingly different from each other. One of these ancient species evolved into all of the arthropods, including the butterfly shown here, but also scorpions, bees, crabs and spiders; and the other evolved into all of the vertebrates – goldfish, cow and platypus (dinosaur, salamander, chicken . . .). Moving up the vertebrate side of our tree, the triangle shows the common ancestor of all the vertebrates. This ancestor is rather intriguing because, while it undoubtedly lived in the sea and looked much more like the goldfish than it did a land-living cow or a platypus, our tree nevertheless tells us that it is precisely as distantly related to the living fish as it is to the living cow or platypus. Finally, we meet the common ancestor of the two mammals shown by the star. This animal was a mammal (hence hairy and with mammary glands); however, like the platypus (and unlike nearly every single other mammal), it laid eggs.

The timescale on my tree is entirely fictitious, starting with the ancestor of these animals, which (according to this tree) existed 500 million years ago, and ending with the animals alive today – so zero years ago. We can use this timescale to convince ourselves that the mammals are no further from the animal common ancestor than the butterfly is, and that the butterfly is equally closely related to the goldfish, cow and platypus. If the timescale were accurate, it would allow us to measure the evolutionary distance between any pair of living species by measuring the total length of the horizontal branches that separate them. This timescale also allows us to work out when common ancestors lived. According to my tree, the most recent common ancestor of the cow and platypus – the mammalian common ancestor – lived about 200 million years ago.* We can also see that however old the vertebrate common ancestor (the triangle) really is, it *has* to be older than the mammalian common ancestor (the star), and the common ancestor of all four animals *has* to be older than any of the species (living or extinct) that descended from it.

One final skill needed to read a tree is understanding that the order of the species from top to bottom, to the extent that it can vary without changing relationships, is arbitrary. In figure 3 I show the same tree relating cow, platypus, goldfish and butterfly

* Having written this I thought I ought to check the real age of the mammalian common ancestor, and I am delighted to tell you that my fictional tree has scored a flukey bullseye: the recently estimated age of the first mammals is between 251 and 165 million years.

but drawn differently. All these trees convey exactly the same information. The only information in the trees is in the branching pattern and in the horizontal branch lengths.

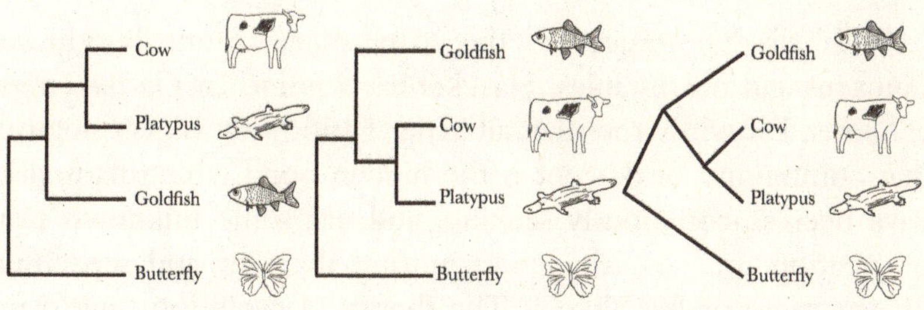

FIGURE 3: Three trees drawn differently but all telling the same story.

We have seen that older classifications of life, such as that of Linnaeus, while inevitably inherently evolutionary (they are classifying the product of evolution), come from a time when the concept of a single origin of life was impossible to imagine. Linnaeus's classification was intended to be useful, but Linnaeus had no conception that it encapsulated the history of life. Haeckel was a brilliant man and seems to have been extremely sure of his own abilities. As a writer he was not one to pull his punches when considering the work of others. In a passage worth quoting at length, we discover his lack of sympathy for earlier classifiers of life such as Linnaeus, considering their efforts almost beneath contempt – the work of stamp collectors.

> Most naturalists who have hitherto occupied themselves with arranging the different systems of animals and plants, have collected, named, and arranged the different species of these natural bodies with much the same interest as antiquarians and ethnographers collect the weapons and utensils of different nations. Many have not risen above the degree of intelligence with which people usually collect, label and arrange crests, stamps and similar curiosities . . . This childlike treatment of systematic Zoology and

> Botany is completely annihilated [*gründlich vernichtet*] by . . . [Darwin's] Theory of Descent. In the place of the superficial . . . we now have the much higher interest of the intelligent understanding which detects in the related forms of organisms their true blood relationships.[12]

With all due respect to the extraordinary contributions of Linnaeus and his disciples, Haeckel has a point. In *On the Origin of Species*, Darwin wrote that all 'true classification is genealogical; that community of descent is the hidden bond which naturalists have been unconsciously seeking, and not some unknown plan of creation, or . . . the mere putting together and separating objects more or less alike'.[13] The theory of evolution transforms the merely useful system of classification of Linnaeus into an extraordinary tool for understanding billions of years of evolution. Evolution provides a single, simple but brilliant explanation for the order found in Linnaeus's system. On top of this, because the order derives from a single, real evolutionary history, there is (unlike library books or stamps) a single perfect ideal to be sought when constructing that classification – a *truth* to be discovered. And, as we will see in the next chapter, the mechanics of evolution naturally churn out the very clues we need to make discovering this truth achievable.

2

The Venus Flytrap and Other Unlikely Relatives

A DOOMED BEETLE CLIMBS up the stalk of a small plant that sits damply among the mosses of a bog. There is plentiful sunlight and unlimited water here, but the lack of key nutrients makes this a tough place to grow, and there's little other plant life around to interest the insect. The plant's form is unremarkable – small, ground-hugging and typically rather unkempt, with a splayed tangle of young, mature and dying leaves; even its white, five-petalled flower is merely pretty. Despite its unpromising form (undistinguished by size, shape or colour), it is a celebrity amongst plants because of some truly unusual traits. Its most surprising qualities only reveal themselves when the poor beetle crawls onto one of its oddly shaped, red-tinged and sweet-smelling leaves. At first, nothing outwardly happens, but tiny hairs on the surface of the leaf have been bent by the weight of the insect, and this has put the whole leaf on high alert. The next time the beetle moves, the next time the hairs are bent, the true character of the funny-looking leaf becomes clear. The leaf is a trap whose two halves now snap suddenly shut, locking the beetle in a tiny prison. The final step is much less hurried – the beetle is going nowhere. Over a period of several days, a chemical soup oozes from the walls of the prison cell to dissolve the body of the beetle. It is the nutrients extracted from the beetle's body – the nitrogen needed to make proteins and the phosphorus needed to make DNA – that are the secret behind the plant's ability to survive in the bog.

Our unremarkable-looking plant is, of course, the famous Venus flytrap, and it is by supplementing a diet of pure sunlight with the flesh of flies, ants, spiders, beetles and the occasional small frog that it has established itself as the unopposed floral king of the marshes. For the purposes of seeing evolution in action, the Venus flytrap is as good an example as any species on the tree of life. We are going to watch as the little part of the tree of life that describes the evolution of the Venus flytrap and three of its closest relatives grows from their common ancestor. We will witness three important aspects of the growth of the tree of life: how new species are created; the evolution in these new species of new characteristics; and the passing down of these novel characteristics to future generations.

The flesh-eating behaviour is something the Venus flytrap shares with just a small handful of other plants – sundews, pitcher plants, bromeliads – and, serving at least to emphasise the innate fascination of this odd way of life, with the fictional Triffids and Audrey II from *Little Shop of Horrors*. Fascinated eighteenth-century botanists described this strange mix of animal and plant as amphibious – living a double life.

The Venus flytrap and the sundew are, of all these carnivorous plants, the most animal-like; not only do they both eat flesh, but they detect their prey and are able to move to capture it. It was these oddest of all aspects of their lives that captivated Charles Darwin who, alongside work on earthworms, barnacles and arrow worms (and helped by his son Francis), studied and experimented on them on and off for fifteen years. In 1860 (immediately after the publication of *On the Origin of Species*), Darwin wrote to the geologist Charles Lyell that 'at this present I care more about *Drosera* [the sundew] than the origin of all the species in the world'.[1] Darwin finally published his findings in 1875 in a book called *Insectivorous Plants*.[2]

The early history of the discovery of the Venus flytrap by European botanists was an undignified business. Europeans first discovered what the indigenous Americans ('either Cherokee or Catabaw but which I cannot now recollect'[3]) called a 'tipitiwitchet' in the 1760s. The word, if the story of its origin is not a complete fabrication, may have been closer to *titipiwitshik*, meaning 'they [i.e. the leaves of the trap] which wind around'.[4] The first specimens that made it to Europe were collected by a 'diligent and indefatigable' botanist, John Bartram, a Quaker from Philadelphia, who found them in 'swamps beyond the blue mountains' (probably in North Carolina).[5] These early botanists were fascinated by the amphibious life of the plants but seem to have been more especially taken with the resemblance they perceived between the plant's trap and female genitalia – a pair of fleshy lobes, reddish in the centre, surrounded by hairs, sensitive to the touch and capable of grasping its prey. The close correspondence between the original word *titipiwitshik* with the existing saucy slang 'tippet-de-witchet' ('wicket' refers to a gate or opening) was too much for them. There is some innuendo in their letters that supports this idea ('my little tipitiwitchet sensitive stimulates laughter in all ye beholders'[6]), but the proof of their puerility is found when one of their number complains that they cannot get flytrap specimens from the recently married Governor of North Carolina: 'It is now in vain to write to him for seeds or plants of Tipitiwitchet now He has gott one of his Own to play with'.[7] The goddess of love is clearly the origin of the name 'Venus', but the first publication to describe and officially name the plant rewrites the lewd narrative. The paper, 'A new sensitive plant discovered', was published by John Ellis in 1768 in the *London Magazine, or, Gentleman's Monthly Intelligencer* (a magazine that should definitely be revived – Ellis's piece was sandwiched between an article 'Remarks on Tooth Powders' and one relating 'The Particulars of the barbarous Murder of the celebrated Abbe Winkelman'). Ellis's fiction has it that 'from the beautiful appearance of its milk-white flowers, and the elegance of its leaves, [Dr Solander] thought it well deserved one of the names of the goddess of Beauty and therefore called it *Dionaea* [that is daughter of Dione, i.e. Venus]'.[8] The full genus and species is *Dionaea muscipula*, the species name being a bit of a fudge between mousetrap (Latin *muscipula*) and flytrap, the Latin for fly being *musca*.

Like all species, the Venus flytrap is unique in some of its characteristics but is nevertheless anchored in its particular place on the tree of life by other features it shares with its nearest and dearest. The evolutionary history of this rather extraordinary plant can be our guide to exploring the question that lies just below the surface of the last chapter: why were the pre-evolutionary classifications of Aristotle, Linnaeus and others so competent? Why do their 'natural' classifications correspond so well to those made later with the knowledge of Darwinian evolution? And why do humans find it easy to recognise that a giraffe and a goldfish belong on one branch of the tree and a seaweed and a gooseberry on another? Let's get stuck into a concrete example by meeting four related carnivorous plants.

The most notable characteristic of the Venus flytrap is, of course, its trap. Evolution has modelled this by enlarging and elaborating a leaf to resemble a bivalved mussel shell fringed with spines; the leaf forms a type of trap called a snap trap. The closest relative of the Venus flytrap is an aquatic plant called the waterwheel (genus *Aldrovanda*, after the Italian naturalist Ulisse Aldrovandi), which has smaller snap traps clustered around the stem, each whorl of traps resembling the buckets on the wheel of a water mill. The closest species to these two snap trap species is the sundew (genus *Drosera* from the Greek word *drosos* meaning 'dewdrops'), named for the glistening, sticky tentacles which cover its leaves. These tentacles – each exuding a sticky droplet – resemble pins in a pin cushion and form a different type of trap called a flypaper trap. The flypaper trap of the sundew nevertheless shares important characteristics with the snap trap; both traps are formed from modified leaves and both are able to sense the touch of an insect and to move in response, capturing the food.

The final carnivorous plant species we will think about – the most distantly related to the Venus flytrap – is the dewy pine (genus *Drosophyllum* – 'dewy leaves'), which for a long time was classified as being most closely related to the sundew. It has the same insect-trapping, dew-tipped tentacles; these cover the whole plant, even the sepals that enclose its little yellow flowers. The dewy pine is larger than the sundew, standing up to 20 centimetres tall, and the living plant is surrounded at ground level by an ugly mess of old, dead leaves. The character that separates dewy pine from the sundew, flytrap and waterwheel is its inability to detect the presence of insects and to move its leaves in response.

Carnivorous plants have a surprisingly long evolutionary history. If we wanted to meet the Venus flytrap's most ancient carnivorous ancestor, we would have to travel some 90 million years into the past, to the Late Cretaceous. This was not so long after flowering plants – the angiosperms – had first emerged, diversified and spread, replacing the gymnosperms (conifers and relatives) as the dominant flora on land. The angiosperms and their flowers evolved, obviously I suppose, hand in hand with insects, their marriage sealed by the giving of nectar and the pollination services received in return. The innovation shared by all four of our carnivorous species – a specially modified leaf that catches and kills insects – turned this gentle relationship between plant and insect on its head. The very first plants of the lineage took the tiny hairs that cover the leaves of almost all flowering plants and enlarged them enormously to make a lawn of 'tentacles'. Each tentacle has a gland at its end which exudes a sticky fluid droplet, and these insect-trapping leaves were the carnivorous plants' key to the bog kingdom.

One of the great ironies of the life of these insect-eating plants is that they still rely on insects to pollinate their flowers – a circle they have squared by holding their flowers at the end of a long stalk, keeping pollinator species as far as possible from the dangers of the traps. An active community of pollinators is essential to facilitate sex between plants. Sex ensures that the genes that exist throughout the species are constantly mixed up, keeping it integrated and relatively homogeneous. It is only by dividing this population that the parts thus separated can start to become different – if one species is to become two, the two parts have to be prevented from having sex. And this process of speciation was essential in turning a single ancestral species into the four that are alive today.

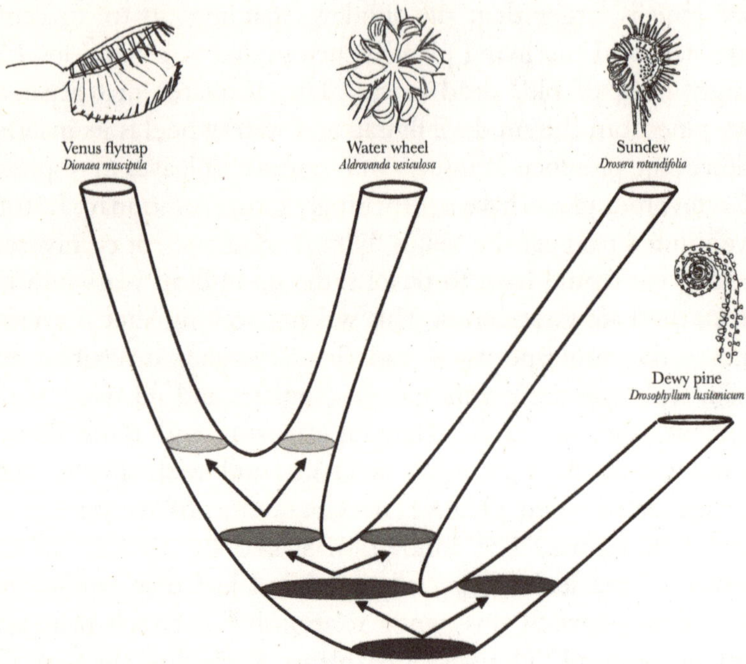

FIGURE 4: Tree relating four species of insectivorous plants showing how populations divide to make new branches.

For simplicity we are going to assume that a mountain range has suddenly risen up in the middle of our population, dividing it in two. Our original single species is now found in two reproductively isolated populations, which are free to evolve to become two distinct species (figure 4). After a few million more years (and a correspondingly long section of the relevant branch), one of those two species divides again, and after a few more tens of millions of years one of *these* two branches splits to make two more. First one species has divided to make two, later on two have become three, and when one of these three finally splits again we end up with four.

This simple account of the evolution of four species has just two components – speciation and time – but even this simple model shows several intimately related things. The tree we have pictured shows us both of these aspects of history: the moments when speciation happened (the places where branches separate) and the time between these separations (the growth of the tree's branches). By tracing the path of the branches that link any two of the living species, we can read the closeness of their relationship and see how long ago their common ancestor lived. We can also see that there is a logical order of speciation events: those separating larger and more inclusive groups *must* have occurred before those separating smaller, more exclusive groups. The split that separated the sundew from the two snap trap species must have occurred before the split that separated the Venus flytrap from the waterwheel.

Our model of speciation plus time is usefully simple but really only gives the barest outline of the rich story of evolution. Our four species are not blank slates whose most exciting attribute is the ability to speciate and to endure. Every species is a complex machine, and each has its own set of observable characteristics – what is called a phenotype (from the Greek *phaínō* – 'to appear' or 'to show'). Not only does each species look like something, but species' phenotypes change over time as a result of evolution by natural selection. There have been countless changes along the branches of our little tree, many of them incredibly subtle but also some that are obvious, such as the changes that led to the hallmark snap traps of the Venus flytrap and waterwheel. Let's add colour to the bare branches of our simple tree by adding some of the characteristics we can find in our four protagonists (figure 5). We begin at the root of the tree, where we find that lots of evolutionary work has already been done: all four species (in addition to having all the many attributes of the angiosperm branch of the tree of life – leaves, flowers, chloroplasts, seeds etc.) have inherited from their most recent common ancestor the ability to attract, capture and digest insects.

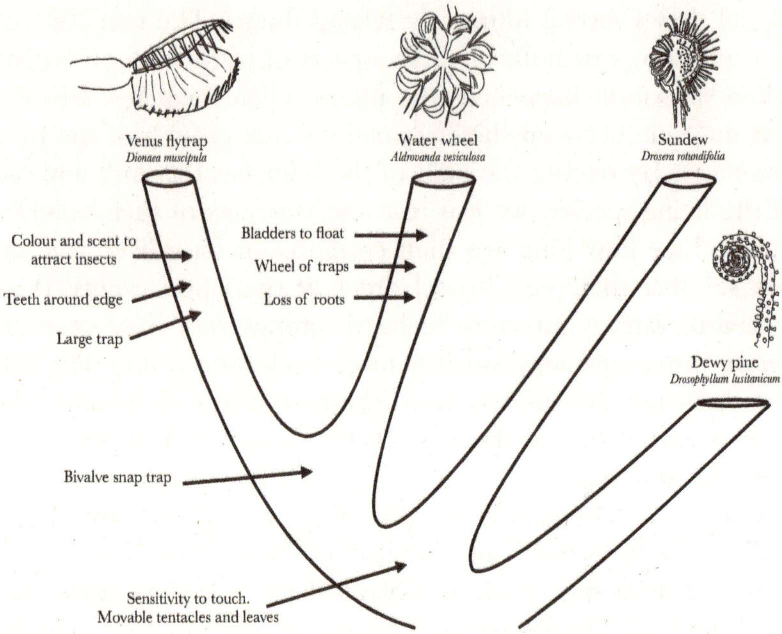

FIGURE 5: Tree relating four species of insectivorous plants showing how new characters are accumulated through time.

Of the two branches separating closest to the bottom of the tree, we are going to ignore the dewy pine (it has served its purpose of showing the characteristics present in the ancestor) and watch what happens to the other branch – the one that leads to sundews, Venus flytraps and waterwheels. Perhaps 70 million years ago (it is very difficult to be sure of dates), the ancestor of these three took the sticky trap that it had inherited and improved it. It evolved the ability to detect the presence of an insect touching the tentacles on its leaves, as well as the ability to move its tentacles and even curl its leaves up and around to trap the poor insect more securely – innovations which are now shared by the sundew, waterwheel and Venus flytrap.* Millions of years later, at some point along the

* This sensitivity to touch is exquisite. Darwin reported as much early in his

branch that leads just to the waterwheel and Venus flytrap, there was an advantage to be found in catching larger insects.[9] The leaf became broader and then began to change shape, becoming bivalved, each side capable of springing shut, and with interlocking spines round its edges to hold the victim in place. The most recent common ancestor of the two snap trap species would pass on to both of its descendants this two-sided trap that was sensitive to the lightest touch of a fly, was able to close rapidly and was ready to digest the tissues of the prey it evolved to accommodate.

Another 10 or 20 million years later, the branches on our tree that would lead, ultimately, to the Venus flytrap and waterwheel would separate from each other, and each would see more changes to produce the different plants we know today. The Venus flytrap has become squatter, and its traps larger, targeting not flying insects so much as the bigger and juicier crawling beetles, ants and spiders. Its trap has developed multiple long, fringing spines and an alluring smell and a deep red colour the better to attract passing insects in search of nectar. The waterwheel has changed even more obviously as it moved from damp land to open water. Its traps have become smaller with fewer spines, better suited to catching water fleas and insect larvae perhaps. No longer anchored in soil, it has dispensed with its superfluous roots entirely and evolved air pockets in its stem that ensure it floats, maximising its exposure to sunlight.

In order to be handed down to future generations, the many changes involved in the evolution of all these new characteristics must be encoded in the DNA of these plants. While we still struggle to predict how changes in DNA (changes to genotype) link to changes in observable characteristics (changes to phenotype), just occasionally the connection turns out to be pretty obvious. For example, some of the genes that code for the growth

sundew studies in an 1860 letter to J. D. Hooker: 'Here is a fact for you, which is as certain as you stand where you are, though you wont believe it, that a bit of Hair 1/78,000th of one grain in weight placed on gland will cause one of the gland-bearing hairs of *Drosera* to curve inwards'. (Darwin Correspondence Project, 'Letter no. 2991' (accessed 11 September 2024).)

of roots in terrestrial plants have been lost from the waterwheel genome – a change that would have been beneficial for a plant that was abandoning the soil.[10] The new rootless condition would now be passed down for generation after generation encoded in the changed DNA. The accuracy of the copying of DNA (geneticists use the word 'fidelity') is extraordinary – the human genome, as a fairly representative example, consists of more than 3 billion individual DNA letters, and when this huge tome is copied and passed from parent to child, only one letter out of every 37 million will have been incorrectly copied. If you were to copy out every volume of the *Encyclopædia Britannica*, this rate of error would result in just eight single letter typos. This fidelity means that changes that appear in an ancestor can be faithfully handed down to descendants over hundreds of millions or even billions of years. It is precisely because characters that evolve in an ancestral lineage (a sticky tentacle, the ability to move or a snap trap) are then carefully passed on to all of the descendants of that ancestor that such characters are the perfect way of identifying those groups of related species.

The consequence of this combined pattern of speciation and the evolution and inheritance of new characters is that closely related organisms (e.g. species within the same genus) share more characters than more distantly related organisms do. Of course this is true, but it is worth unpicking the reason why. Species that belong to the same kingdom will share characteristics common to all members of that kingdom. Two animals, for instance, will have inherited all the characters that belonged to the last common animal ancestor (e.g. DNA in a nucleus, having multiple cells, moving with muscles, a gut, some kind of a nervous system). A snail and a whale – both in the animal kingdom but in different phyla, and thus very distantly related – share these universal animal characters but no others. The branches that lead to snails and whales parted a little over half a billion years ago, when the last common ancestor lived, but have evolved independently ever since. Contrast this with two species that are related at the level of the genus, such as a lion and a tiger; these two will share the characters of the kingdom that they both belong to, but also those of their shared phylum (vertebrates), class (mammals), order (carnivores), family (cats) and genus (*Panthera*). The result, of course, is that two species in the same cat genus have a great many characters in common. Two closely related big cats look very, very similar, and this similarity is written in their DNA.

A slightly less obvious implication, perhaps, is that the various characteristics themselves have both different ages and correspondingly broader and narrower distributions across the tree of life. Characters that define kingdoms evolved long before characters that define genera, and kingdom-defining characters are found in lots of species (in all the many members of the kingdom) while genus-defining characters are found only in the handful of species in that genus. To apply that to us, our animal kingdom characteristic of multicellularity (a body made of many cooperating cells) evolved in the Precambrian, perhaps 600 million years ago, and is common to all 8 million or so species of animal (including, of course, snails and whales). The human ability to speak evolved in the last few million years and is found in just a single living species.

With all of this in mind, we are now in a position to answer the question we asked at beginning of this chapter of why we (and, more pertinently, Aristotle and Linnaeus) are so easily able to classify organisms in a way that corresponds to their evolutionary relationships. The branching pattern of relationships coupled with the evolution of novel characteristics in ancestral species, and the inheritance of these new characters by their descendants, is the reason why the natural classifications of Aristotle and Linnaeus, which were based only on what they could observe, end up reflecting the evolutionary tree of life. Close relatives share lots of characters. More distantly related branches are more dissimilar, and because Aristotle and Linnaeus based their classifications on characters *that behave in the way described* – changing occasionally, and these changes getting passed on down the generations – these two brilliant scientists inevitably came up with classifications that reflected the evolutionary history of life.

One thing you may have noticed is that our tree has effectively allowed us to travel back in time. We can read the tree to discover when characters first appeared and in this way catch a glimpse of the phenotype of the long-dead common ancestor of sundews, waterwheels and Venus flytraps, a species which we can never see in the flesh. These ancestors are anonymous organisms, represented by a moment on the tree of life where two branches separated, but there are other ancestors whose existence is much more tangible. They are found in the fossil record. How we can incorporate these fossils into the tree of life – they are as much a part of it as anything alive today – and what they can tell us about the history of life on earth, is where we will venture next.

3

A Distant Cousin from the Ocean's Depths

IN 1978, A long series of human-like footprints were found fossilised, alongside the footprints and fossilised remains of numerous other species of mammals, birds, reptiles, molluscs and insects, in a layer of volcanic ash in a place called Laetoli, close to the Serengeti National Park in Tanzania.[1] The layers of ash immediately above the footprints were dated to 3.66 million years ago, meaning the footprints must have been made by something (someone?) alive just before this distant point in time. The footprints come from three individuals walking side by side across a fresh fall of ash, and, while they show a slightly different gait from modern humans, they were clearly walking upright, on two legs. They were made by human relatives, members of the species *Australopithecus afarensis* known from contemporaneous fossils found in the region (including the famous Lucy).

These *Australopithecus* fossils take us to a place on the tree of life that we could never visit by studying only the species that survive today. We are bipedal, but our closest *living* relatives, the chimpanzees, are not. From the sparse information that we can extract from a comparison of living humans and chimps, we can know for certain that bipedalism evolved at some point in the 8 million years since the human line parted company from the chimps, but we have no way of narrowing down when in this huge time period this happened. The fossilised footprints of Laetoli shine a light on this otherwise murky 8-million-year period; they are direct evidence that the human ability to walk on two legs had already evolved 3.6 million years ago.

This snapshot of the life lived by these distant human relatives, and specifically the evidence it brings of a characteristic shared with our own species, is a handy reminder that, like these ancient, two-legged hominins, almost all the diversity of life that has ever existed is now extinct. The millions of species alive today are a tiny fraction of the billions of species that have ever populated the earth.

From the point of view of how we read the tree of life, almost everything that is true for living species is also true for species we know only from their fossils; and fossils can be straightforwardly included on our tree of life. Every fossil was formed by an individual of a species that died a long time ago. Its twig does not reach the top of the tree of life but terminates lower, marking its demise at some time in the past. Placing a fossil species on a tree shows the exact same information as for living species: its relationships to other species and the span of its existence on earth. Furthermore, fossils are included in the exact same classification as living species – a fossil trilobite is a member of the arthropod phylum, an ammonite is a mollusc, a giant sloth is a mammal. And, finally, just like living species, fossils have the most characters in common with the species, living or dead, that they are most closely related to.

Fossils of extinct species, besides being dead, differ from the living in other important respects, most obviously in what is and what (usually) is not preserved in a fossil. Most important – and this is true for the vast majority of fossils – we have no access to their DNA. This is because DNA, along with all soft tissues, degrades rapidly. Despite the promises of *Jurassic Park*, the oldest intact DNA found to date is 2 million years old; amazingly old, but dinosaurs had been extinct for 63 million years at that point. To put it in perspective, the 2 million years that DNA might just survive covers only the last 0.3% of the history of the animal kingdom and just the final 0.05% of the total history of life. DNA is an incredibly useful tool for building the tree of life, but for almost all extinct species we must rely on their characters that do endure to place them on the tree of life.

It is not surprising that hard parts like bones and especially teeth are much, much more likely to remain intact long enough to produce a fossil than soft tissues. Somewhat disappointingly, almost the entire fossil record of mammals is composed of teeth. With wonderful exceptions like the Laetoli footprints, fossilised records of movement, colour and behaviour are even rarer than fossilised flesh, giving us a very incomplete picture of the biology of extinct species. Just occasionally, however, we are able to make this leap back in time via living species that seem frozen in time. Ancient beings that live amongst us like a caveman in the crowds of Trafalgar Square.

Our example tree of carnivorous plants showed species splitting and time passing, and I have so far assumed that the lengths of the branches we see on a tree of life tell us how much time has passed. I have to admit that I have been slightly economical with the truth. Often what the length of a branch actually shows is how much change has happened. Usually, the number of changes is closely correlated with how much time has passed – more years means more changes – and so branch length indicates roughly how much time has passed. But sometimes this correlation breaks down, and it is here – in the exceptions to the rule – that we find some of the most interesting stories of evolution.

In 1936, chance introduced two people from very different worlds – Captain Hendrick Goosen, skipper of the fishing trawler *Aristea*, and Miss Marjorie Courtenay-Latimer, the twenty-nine-year-old curator of the small East London Museum in South Africa. They met, improbably, on Bird Island, forty miles east of Port Elizabeth (today called by its Xhosa name of Gqeberha), where Courtenay-Latimer was spending a few months collecting specimens and where Goosen frequently stopped to collect rabbits to feed his crew. The two struck up a friendship, sealed when Goosen agreed to ship Courtenay-Latimer's crates of specimens back to the mainland. Goosen also undertook to bring back any interesting-looking fish specimens that his crew trawled up for the museum's new aquaria, and a container was installed on his boat to keep them alive. Two years later, on the morning of 22 December 1938, Courtenay-Latimer was called to sort through a ton and a half of specimens that had just come in on Goosen's boat. The first look through showed many familiar (and unwanted) specimens, but then she noticed a blue fin sticking up from beneath the pile. Uncovering the roughly 5-foot-long specimen, she describes the emergence of a pale blue fish: 'the most beautiful fish I had ever seen . . . with faint flecks of whitish spots; it had an iridescent silver-blue-green sheen all over.' After a little trouble with a taxi driver reluctant to transport the smelly fish, she was able to get it back to the museum wrapped in a sack.

She had never seen anything like it, but even then, she had an inkling as to what the fish was, and what it might mean. Consulting her limited library, the best match, to the long-extinct coelacanthid fishes, was surely wrong. These fish were well known, but only from their 400-million-year-old fossils. 'This is impossible . . . This is a live fish! It cannot be a fossil fish!' Courtenay-Latimer now struggled to get an expert to examine it; it was Christmas, and, as the fish began to decompose, a lack of enough formalin to preserve it whole obliged her to have it stuffed. Finally, a whole month and a half later, Courtenay-Latimer showed it to an amateur ichthyologist colleague, James Leonard Brierley Smith, who immediately confirmed the connection she had made with its almost identical 400-million-year-old relatives. Her diary records his reaction: 'Lass, this discovery will be on the lips of every scientist in the world.'[2] And he was not far wrong.

Rates of morphological (and, less often, genetic) change can lose their correlation with time either by slowing down or by speeding up. Unusually rapid change can happen in a species when a new opportunity presents itself – such as arriving on a deserted island with no predators or with new food sources – or when a new challenge appears, such as a predator inventing a new way of eating you. Rapid change is of special interest from the very practical point of view of someone (like me) who wants to reconstruct the tree of life correctly, because it almost always adds confusion.

The second way the correlation can be lost is when morphological change slows to become barely perceptible over huge periods of time. Groups of organisms that have barely changed for hundreds of millions of years are famous as 'living fossils' (the quotation marks highlighting this as a slightly disreputable concept).

The strange fish discovered by Courtenay-Latimer is perhaps the best known of all the 'living fossils'. The coelacanth is a member of a group of bony fishes called Sarcopterygii, which are well known from the fossil record from the Early Devonian, 410 million years ago,[3] right through to the end of the Cretaceous, 65 million years ago, at which point they disappear along with the dinosaurs. 'Sarcopterygii' is usually rendered as 'lobe-finned fishes' (though the proper translation of the ancient Greek is closer to 'flesh-finned'). But, surprising though it may seem, all vertebrates with four legs – amphibians, reptiles (including the feathered dinosaurs we call birds) and mammals (including, of course, humans) – are also members of this group. The precursors of our own limb bones – femur, tibia and fibula of the leg; humerus, radius and ulna of the arm – can be recognised in the fleshy fins of the sarcopterygians. The coelacanth, while undoubtedly a fish, is more closely related to a human than to a goldfish (which lacks these fin bones). A large part of our fascination with this species derives from this pivotal role that coelacanths seem to have in our own origins story.

Smith published the first description of the living coelacanth in the prestigious British science journal *Nature*, in an article that begins with Aristotle's phrase (in Pliny's Latin) *Ex Africa semper aliquid novi* – 'Out of Africa new things always come'.[4] He placed the fish in a new genus named *Latimeria* (to honour Ms Courtenay-Latimer) and gave it the species name *chalumnae* (after the Chalumna River which enters the sea close to where it was caught).

Latimeria chalumnae has since been found in several places around the western Indian Ocean between Madagascar and Mozambique, and a second coelacanth species in deep waters around Indonesia. I had a brush with one specimen in Madagascar in the summer of 1988. I was an undergraduate on an expedition studying the endangered side-necked turtle. All the clichés of the tropics applied, starting with the praying mantis that landed on me outside the airport terminal. Elsewhere, we saw other members of Madagascar's famously unique flora and fauna, including lemurs, chameleons, travellers' palms, baobabs and tapia trees. More toe-curling memories of the expedition include my decision not to wash my hair for three months 'because it would clean itself' and my pathetic attempt to grow a beard. At the end of three months, we counted ourselves lucky to see three specimens of the side-necked turtle, caught (and then eaten) by the fishermen who looked after us on a tiny island in the middle of Lake Kinkony. I saw my specimen of *Latimeria chalumnae* in an outside corridor of the biology department of Antananarivo University. It was unforgettably and spectacularly sad, a superstar of evolutionary biology lying incognito, pickled in the murky liquid of a very dilapidated fish tank. Students rushed past it to their classes with no idea what they were missing. There is a surprising and happy end to this story because, to my delight, I have been able to trace the fate of my friend (and distant relative). This same fish now resides, dried and stuffed, in a splendid display case in the Museo Civico di Storia Naturale in Comiso, on the Italian island of Sicily – you can go and see it!

All living sarcopterygians – coelacanths, mammals, reptiles, amphibians – are separated from their common (Devonian) ancestor by precisely the same number of years. Their branches all start at this one place on the tree. But if we use the degree of change to decide the lengths of their branches, the branch leading from the ancestor to the coelacanth would be very much shorter than the branches leading to any of the other descendants. The coelacanth has barely changed; the others (frogs, elephants, pheasants and humans) have become radically different from the ancestor. The glacial rate of change is almost certainly telling us that coelacanths are beautifully adapted to their unusual, long, lugubrious lives, spent hundreds of feet below the surface of the ocean. They have been lucky enough to live in such a stable environment that natural selection has acted, via a process called stabilising selection, to ensure that they have been preserved in this ideal state. The case of the coelacanth shows us that the degree to which two species appear different from one another is not always a reliable indicator of how distantly related they are.

I want to end this chapter with a warning about what trees can and cannot tell us about the past. The coelacanth, unchanged as it seems to be, seduces us to imagine that it shows us, in a living form, what our sarcopterygian ancestor looked like. Could the coelacanth *Latimeria chalumnae* be a living representative of the lobe-finned fish stage of human evolutionary history? This idea that we evolved *from* the coelacanth seems a reasonable proposition considering the existence of coelacanth-like fossils (but not human-like fossils) in 400-million-year-old rocks. This is an understandable misapprehension, because we certainly evolved from something resembling a modern-day coelacanth. The problem with this way of thinking arises when we are tempted into imagining that we can always use living species as stand-ins for the stages we passed through on the way to becoming humans. To see the folly of this, think about a series of more and more distant living relatives of humans. Our closest relatives are the chimpanzees; slightly more distant are the lemurs. So far it doesn't seem too far-fetched to think that in our evolutionary history we passed through a lemur stage then a chimp stage before becoming human. The mistake reveals itself as we go to more distant relatives. Slightly more distantly related are whales (we are both placental mammals), then come duck-billed platypuses, then birds, then frogs, coelacanths, goldfish, starfish, fruit flies, jellyfish, mushrooms and so on. I hope the point is clear that we did not in our evolutionary history pass through a mushroom stage before becoming a jellyfish then an insect then a starfish then goldfish, and so on. For none of our evolutionary history have we looked like a chicken or a whale. The series of evolutionary stages through which we have actually passed, unlike the increasingly distantly related living branches listed above, are all extinct species.

What *is* true is that, working backwards, we share a relatively recent common ancestor (and have many characters in common) with a lemur, and a bit earlier than that in our history we have a common ancestor with a whale (slightly fewer characters in common), and before that with a chicken etc. In every case *both* the branches splitting from the common ancestor have – unlike the coelacanth – changed dramatically since their split. The common ancestor that humans share with chickens lived about 320 million years ago and looked neither like a human nor like a chicken.

Simply lining up increasingly distant relatives doesn't let us piece together the long series of events that led to our own evolution, but there is, luckily, an approach that does work. This method requires us to think not about our living relatives themselves but about the characters we share with them and what these can tell us about our ancestors. I suggest that it is the evolution of characters, perhaps even more than species, that evolutionary biologists really ought to be interested in. This character-focussed way of thinking is a route to asking some big questions. When did the backbone or feathers or leaves or multicellularity appear? What came first, the chicken or the egg? What characteristics have appeared and then been lost (snakes' legs, the human tail), when did these things happen and why? All these questions and many others rely on deciding where on the tree of life a character evolved. There is a logical and (mostly) reliable way to discover this, and the simple trick, a trick we will also use to build the tree of life itself, is the subject of the next chapter.

4
The Real History of the Birds and the Bees

THE LIFE OF Dmitri Mendeleev reads like a Dostoyevsky novel. It started well: he was born in Siberia in 1834 to parents who were educated and reasonably well off, his father a school principal and his mother (his poor mother who produced seventeen children, of whom he was the last) coming from a family of merchants. But the respectable, comfortable family life did not last; his father went blind and lost his job, so his mother, to support the family, opened a glass factory, but this burned down, and then, when Dmitri was just thirteen, his father died. His extraordinary mother was determined for her youngest son to succeed and, two years into widowhood, took the fifteen-year-old Dmitri 2,000 miles across Russia to Moscow and then – when the university in Moscow didn't accept him – to the university in St Petersburg. A year later, both his mother and his youngest sister (who had made the journey too) were dead from tuberculosis. Mendeleev survived and married and then, having threatened suicide if rejected, he married again, bigamously. He died at the age of seventy-two following happier years of fame, respectability and enormous international recognition for his scientific brilliance. Mendeleev's great achievement was the periodic table of the elements, which he constructed using the regularly (periodically) changing characteristics of the elements when they were arranged according to their mass. The periodic table is much more than a simple classification of elements, however. Its superpower lies in its ability to make predictions about the elements and to explain their chemistry.

When Mendeleev placed all the known elements in his table, he noted gaps where an element with a mass halfway between two known elements should have existed. Mendeleev's extraordinarily brave next step was to predict not only the *existence* of these entirely unknown elements but also their *characteristics* – their atomic mass, density, melting point, valency, reactivity, conductivity, malleability etc. – these properties being predictable from their places in the table. He named four of these predicted elements ekasilicon, ekaaluminium, ekaboron and ekamanganese, the prefix *eka* (from Sanskrit meaning 'one' or 'first') indicating they were one row down in his table from silicon, aluminium and so on, and expected to share their characteristics. Mendeleev's fame was guaranteed when the first three of these, possessing chemical characteristics remarkably similar to those he had predicted, were discovered and named – respectively – germanium, gallium and scandium.*

The periodic table has changed little since Mendeleev, bulging and extending to accommodate new and heavier elements but retaining all the principal features of the original. The table cannot change; it is essentially unimprovable as a way of arranging the elements according to their chemical characteristics, simply a representation of an underlying truth about the nature of elements. There is a fundamental reason for the sharing of chemical properties: the number and arrangement of protons, electrons and neutrons in the atom, and the physics that lies behind the single possible arrangement of elements in the table, ultimately explains much of chemistry. All of this future knowledge was hiding latent in Mendeleev's table, waiting to be discovered by twentieth-century physicists and chemists. This immanent predictive power is an attribute that this icon of chemistry shares with its biological counterpart – the tree of life.

Like Mendeleev's periodic table, there is just one true tree of life waiting to be discovered. And like Mendeleev's table, the tree

* Ekamanganese (now named technetium) is an unstable radioactive element not naturally found on earth and was eventually discovered in 1937 in a sliver of molybdenum foil that had been bombarded by deuterium nuclei inside a cyclotron.

of life contains huge power to predict and to explain. The tree of life is the key to understanding the evolutionary history of all parts of biology – the origins of genes, cells, behaviour, embryology, morphology, biochemistry, eating, breathing, reproducing and everything else that we know of the complexity and diversity of life today. But before we can begin to build the tree of life, we must deal with a series of traps that await us and find ways to avoid them.

I love watching swifts and swallows hunting for insects. I am lucky that, in the middle of the part of London where I live, there are plenty of swifts to be seen (and their screams heard) on summer evenings. And when I was growing up in a country backwater in Wiltshire, there was a nearby stately home (at that time it was a private school that I earned money painting and decorating in the holidays) where swallows nested high up under the eaves. They would plunge from their nests, diving straight down, levelling out just above the long grass where the insects were to be hoovered up. Both the shapes of their bodies and their flying abilities *sans pareil* reminded me of spitfires. Swifts and swallows look and behave remarkably similarly, and they resemble each other in other ways too. Both species like to nest high up under the eaves, and both migrate with the seasons, returning to their nests in spring.

Where they go when they disappear in the winter was a puzzle for many centuries. In a charming but dubious story, in late summer the twelfth-century monk Caesarius of Heisterbach is said to have attached to the leg of one bird a parchment note that asked, 'Oh swallow, where do you live in winter?' When the bird returned to its nest in the monastery the next spring it carried a note in reply which read, 'In Asia, in the home of Petrus', thought to refer to Palestine, the home of St Peter. Other ideas were that swifts and swallows wintered on the moon or, slightly more plausibly (and indeed most popularly at that time), that they hibernated underwater, buried in mud at the bottom of a pond like a frog. This idea was accepted as at least as plausible as their flying to Africa by the great naturalist Gilbert White.* Evidence against this notion can be found in a paper published in 1823 by the (at that point recently deceased) Dr Edward Jenner, famous as the inventor of vaccination. Jenner (who on the whole seems like a very decent man) conducted an experiment in which he tested the theory by holding a poor swift underwater: 'I have taken a swift . . . and plunged it into water; but like the generality of animals which respire atmospheric air, it was dead in two minutes.'[1] A successful experiment with a nice clear-cut result but one which I imagine (I hope) brought mixed feelings to Jenner.

As well as having similar migratory patterns, both swifts and swallows (and the closely related martins) are beautifully built for fast, acrobatic and effortless flying: light, streamlined bodies; long, curved wings; large, acute eyes; longish tail feathers (extra-long in male swallows due to sexual selection); small feet (almost invisible in swifts); and short, very broad beaks, giving a wide gape so they can hoover up insects on the wing.

I have to confess that the almost identical forms of the body and behaviour of the swifts and the swallows meant that, until recently, I had assumed they were closely related on the tree of life. If true, this would tell us that we would not have to travel too far into the past to find their common ancestor – perhaps a few million years. I was spectacularly wrong – swallows and swifts are as distantly related to one another as a hummingbird is to an owl, and we have to dive at least 75 million years into the past to find their branching point. My misconception puts

* 'But what struck me most was, that, from the time they began to congregate, forsaking the chimnies and houses, they roosted every night in the osier-beds of the aits of that river. Now this resorting towards that element, at that season of the year, seems to give some countenance to the northern opinion (strange as it is) of their retiring under water. A Swedish naturalist is so much persuaded of that fact, that he talks, in his calendar of Flora, as familiarly of the swallow's going under water in the beginning of September, as he would of his poultry going to roost a little before sunset.' (White, G. (1789). *The natural history and antiquities of Selborne, in the county of Southampton: With engravings, and an appendix*. B. White & Son, London.)

me in excellent company: Linnaeus grouped them in the same genus, *Hirundo* (the Latin for 'swallow'), naming the swallow *Hirundo rustica* ('swallow of the countryside') and the swift *Hirundo apus* ('swallow with no feet' – the swift's feet are indeed tiny, presumably because very rarely used, the swift spending the great majority of its life in continual flight, day and night[2]). The true relationship between swallows and swifts only became clear in the early nineteenth century as a result of the great expansion of comparative anatomy – the distinction relied not on easily appreciated similarities but on detailed analysis of their skeletons (the details – 'osseous bridge from transverse process to caudal articular process on third cervical vertebra; acrocoracoid process of coracoid unhooked' – are terrifyingly esoteric[3]). Swifts, it turns out, are closely related to hummingbirds and nightjars (part of a branch called the Caprimulgimorphae – 'goat suckers'!), while swallows are members of the songbirds (Passeriformes – 'sparrow-shaped') and so are closely related to crows, tits and wrens.

It follows from this that the external characters shared by swifts and swallows – wide beak, curved wings, large eyes, small legs, flight abilities, insect diet and migration behaviours – must have evolved twice, once on the branch leading to swifts and again on the branch leading to swallows. This process, called convergent evolution – most evolution is divergent (species become less similar) – occurs when distantly related organisms come up with similar ways of solving the same problems. For example, long, legless bodies have evolved time and again in the reptiles and amphibians: separately in snakes, slow worms, snake lizards, blind skinks and glass snakes; in both of the little-known amphibian groups of caecilians and sirens; and again in the 300-million-year-old fossil *Phlegethontia longissima*, whose long body might be deduced from its species name. Meanwhile, dolphins, manatees and extinct ichthyosaurs all independently evolved similar aquatic bodies, starting from three different land-living ancestors, each turning what began as legs for walking into fins for swimming. The camera-like eye of humans and other vertebrates finds its close equivalent in the eyes of squids and octopuses, and so on right across the tree of life, with convergent evolution found everywhere from the shapes of bodies, or leaves, or seeds, or limbs right down to the order of the amino acids that make up a protein.

It was clearly an error to use these convergently evolved characters (as Linnaeus and I mistakenly did) to link swallows and swifts on the tree of life, but this kind of error is a really useful clue as to the sort of character we *should* be looking at when trying to link species. A useful character, an honest one that tells us the truth about two species' relationship, would be one that they share because they have both inherited it from their common ancestor. These useful characters are called homologs, and 'homology' is one of the most important ideas in evolutionary biology.

A creationist view of the origin of species doesn't have (cannot have!) any concept of common ancestors nor, therefore, of characters shared by two species that have a common origin. The idea that two species might have characteristics in common that were in some special way the *same* is, nevertheless, much older than *On the Origin of Species*. Aristotle recognised many characters in various animals that he felt need not be discussed individually because they were fundamentally equivalent, even if not identical, like the wings of a heron, a wren and an owl.

A more organised and explicit comparison of equivalent characters between species, specifically of vertebrate skeletons, can be found in *L'histoire de la nature des oyseaux* (1555) by Pierre Belon, where he shows, in two adjacent figures, the skeletons of a man and a bird with their equivalent bones labelled and remarked upon.[4] Belon notes the exact equivalences, despite their very obvious differences in overall shape and function, of human arms and bird wings. These homologs, as they became known, contrast with analogs which, while superficially similar (like birds' and insects' wings), have no fundamental similarity.

Homology has a very special significance both for correctly

building the tree of life and for making sense of it. It is a word that was first defined in print by Professor Sir Richard Owen (founder of the Natural History Museum in London) in his book *Lectures on Comparative Anatomy* (1843).[5] Owen was an unusual-looking man, with a huge forehead, long limp hair and a whiff of Marty Feldman in *Young Frankenstein* about him. He published breathtakingly original work on diverse groups of living and fossil vertebrates; amongst his many achievements was the coining of the name 'dinosaur'. A hands-on comparative anatomist, Owen had been granted first refusal on any animal that died at London Zoo – on one notable occasion, his wife came home to find a dead rhinoceros in their front garden.

Owen was, to put it euphemistically, a bit tricky, and seems to have expended a lot of effort in promoting himself, putting down others and generally taking credit where it was not due. In an anonymous review of *On the Origin of Species* he criticised Darwin's ideas:

> These [observations on pigeons, bees and ants] are the most important original observations, recorded in the volume of 1859: they are, in our estimation, its real gems, – few indeed and far apart, and leaving the determination of the origin of species very nearly where the author found it.[6]

He went on (still anonymously) to refer repeatedly to the marvellous work of a certain Richard Owen. He had a bitter dispute with T. H. Huxley, 'Darwin's Bulldog', over the unique composition of the human brain when compared to other apes, and indeed Owen's classification of mammals placed humans in their own subclass (due entirely to the supposed difference in their brains), implying that humans are no more closely related to other primates than they are to bats or kangaroos.[7] His poor behaviour led the normally mild-mannered Charles Darwin to complain, 'Upon my life I am so sorry for Owen; he will be so d–d savage; for credit given to any other man, I strongly suspect, is in his eyes so much credit robbed from him.'[8]

In *Lectures on Comparative Anatomy*, Owen doesn't feel the need to give a careful and extended definition of homology. He uses the term freely, as if it were already in common usage, and only ventures a terse definition in the book's glossary: 'The same organ in different animals under every variety of form and function'.[9]

Owen and other nineteenth-century comparative anatomists were not Darwinists. Their concept of homology was not of a character inherited by two different species from a common ancestor, but simply of an archetypal aspect of nature that dictated how animals (and indeed any other living things) were built. Owen explains this in an 1849 book about vertebrate limbs, saying that homology signifies 'that essential character of a part which belongs to it in its relation to a predetermined pattern answering to the "idea" of the Archetypal World in the Platonic cosmogony'.[10] One of Owen's aims was, by looking at all the variants of a given homolog (e.g. all the different forelimbs of vertebrates), to infer what the archetypal vertebrate forelimb looks like. This evolution-free version of homology has been called, for its Platonic underpinnings, 'idealistic homology'.

The implication of Darwinian evolution – that all organisms have common descent and that characteristics are passed on (more or less altered) from ancestors to descendent species – gives a very different explanation of homology ('the same organ in different animals under every variety of form and function'). The obvious correspondences between the bones in the wing of a bird and the arm of a man described by Pierre Belon have not arisen because they were each independently adapted from some metaphysical universal template.

The extensive similarities between arms and wings exist because both have been inherited from a common ancestor that itself had a forelimb (which was not yet either an arm or a wing but a precursor of both). We can take this an important step further: every species with a homolog of this limb must (by the definition of homology) be part of the branch of the tree of life that grew from the creature that first evolved the limb. Contrariwise, any creature (a fish, an earthworm, a fly) that *doesn't* have a homolog of this front limb (with all of its bones) cannot belong to this branch – cannot have descended from this legged ancestor – and so must be a more distant relative.

The simple observation that more closely related species share more homologs means that if we want to work out how three species, 'a', 'b' and 'c', are related to each other (the simplest possible problem), we can look for homologs shared between each of them. The more closely related pair of species, let's say 'a' and 'b', will have more homologs in common with each other than either will have with 'c'. This intuitive approach can be scaled up to any number of species: for any given set of species (a, b, c, d, e, f . . .), if we can identify what each species has from a list of homologs (backbone, hair, wings, milk, feathers . . .) then we can work out how closely or distantly related they are to one another.

As we have briefly seen, homologs are the raw data produced by evolution that we can use to work out how species are related to one another. But homologs are also an interesting problem in themselves: how can we look at structures in two species – organs, appendages, teeth, ornaments – and know whether they are homologous rather than, like the wings of birds and bees or the general form of swifts and swallows, just similar? And once we have identified a homolog, what can it tell us about the ancestor of all the living species that possess it?

If two species (for example, pigeons and humans) possess a homologous structure (an arm/wing), then by the definition of evolutionary homology that structure must have been present in their common ancestor. In the case of pigeons and humans, this ancestor existed in the late Carboniferous – a time of rising temperatures, colliding land masses and huge equatorial swamps, when plants would be buried by occasionally rising seas to leave massive deposits of coal (which give the period its name). Thanks to homologs, we know much more about this ancestor than that it had limbs; we can be confident that it possessed every homologous character shared by its bird and mammal descendants. Many of these characteristics turn out to be linked to living more easily on land: it had enhanced its ability to breathe by moving its ribs as well as its diaphragm; it had a new bone in its feet called an astragalus that gave it an ankle; and, most important of all, it laid the desiccation-resistant amniote egg still found in reptiles, birds and (hanging on by the skin of its teeth) the mammalian duck-billed platypuses and echidnas.

This way of thinking about homologs uses a simple but important logical trick called the parsimony principle. This is a common method used across science and philosophy to choose between different models or theories that might explain how the data you have collected were generated. It is an old idea used, inevitably, by Aristotle, whose formulation can be stated as: 'All other things being equal, we should prefer the model that makes the fewest assumptions.' In other words, we ought to prefer a simpler explanation of any phenomenon rather than one that invokes complicated or unlikely factors. The parsimony principle is commonly called Occam's razor after the late Middle Ages monk and philosopher William of Occam (or Ockham), who was a keen proponent – the image of the razor representing the cutting away of superfluous parts of a model.

It is probably easiest to understand both what I mean by a model and how we can use Occam's razor with a simple example. We are going to look at a very simple tree that relates a seagull, a human and a goldfish, and we are going to take for granted that we already know how these three are related – that the human and bird are each other's closest relative and that the fish is more distantly related to both. We will use the parsimony principle to find out when a character (in this case legs) evolved, and in doing so discover whether human and bird legs are homologs.

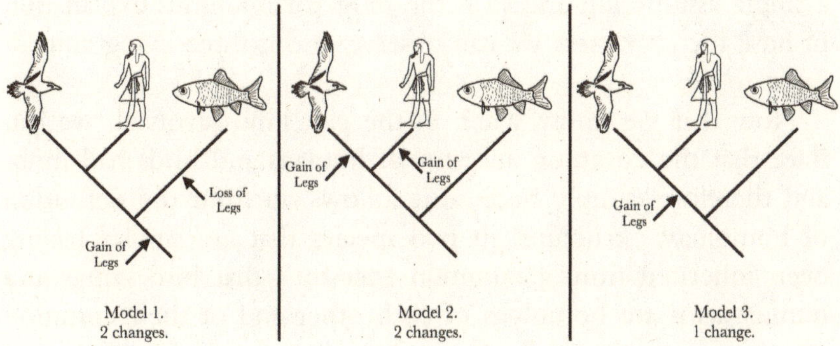

FIGURE 6: Three models of how legs evolved. Model 3 is the most parsimonious and most likely.

There are three different models that could explain how birds and humans have ended up with limbs and fish without; our task is to pick the simplest – the one that makes the fewest assumptions. Model 1 assumes that limbs evolved in the ancestor of all three species, at the bottom of the tree. But, because fish do not have limbs, we are obliged to make a second assumption which is that fish then lost their limbs. For model 1, therefore, we must make a total of two assumptions. Model 2 would consider that the fish evolved from an ancestor that had no limbs (no assumptions made so far) but then we are going to hypothesise that limbs evolved twice later on – once along the branch leading to the human and a second time, independently, along the branch leading to the bird. In this model, the limbs of the bird and the human are *not* homologous, having evolved convergently rather than having been inherited by both from a common ancestor. Like model 1, model 2 requires us to make a total of two assumptions. Model 3 assumes that limbs evolved a single time in the branch leading to the common ancestor of the bird and the human and was then inherited by both of these species from this common ancestor; this model is of course the simplest of the three. It should be pretty obvious that we should prefer model 3; it makes a single assumption and so is the most parsimonious explanation of how the characters we can observe in our three living animals came to be.

Now that we know when in the past limbs evolved, we can infer that the common ancestor of humans and birds had limbs and therefore (simply because it follows on from the definition of homology, 'structures in two species that are similar having been inherited from a common ancestor') that bird wings and human arms are homologs of each other and of their common ancestor's primitive limb. From this triumvirate of homologous limbs (of birds, humans and their ancestor), we have been able to discover something about the evolution of limbs that is really useful in our grand quest to know the events of evolution. We now know roughly what the starting point of human and bird limb evolution looked like, as well as both end points.

There are further nuances in this story that can give us extra help when trying to identify homologous characters. The first is that characters in two species that are similar in complex ways are more likely to be homologous than those with superficial similarity. This is beautifully apparent when we look at bird and human forelimbs, where the similarity goes way beyond the simple fact that both are things sticking out at the front end of an animal. As Pierre Belon noticed, the similarities are legion, with a correspondence of bone after bone. Richard Owen (in *On the Nature of Limbs*) lists these bones as follows:

> The arm proper is appended to this [scapular] arch: its first joint or segment is formed by a single long bone, the 'humerus'; its second joint, by . . . the 'radius and ulna'; and the hand or third segment is formed by a group of little thick bones, the 'carpals', and by five rays or digits; one (i) consisting of three segments, the rest (ii–v) of four segments each; the five bones joining the carpus being called 'metacarpals', and the others the 'phalanges'.

A bird wing, with its entirely different function, has almost every single one of these bones, and in each case the bones are found in the same positions relative to one another.

If we think about these hugely complex correspondences between arm and wing in terms of parsimony, a model that requires these two appendages (with all the similarities we have seen) to have evolved twice over will obviously be hopelessly less parsimonious than a model that tells us that they evolved once. The number of evolutionary coincidences we would have to propose is multiplied by the number of similarities. To put it simply, two structures that have many complex similarities are much more likely to be homologous than two simple structures that would be trivial to evolve. Two Picasso self-portraits share such complex similarities (subject, style, brushwork, colours, composition) that they can only have been made by the same creator; two paintings that consist of a single red dot, in contrast, could have been done by anyone.

This important idea can be approached from the other direction entirely by thinking about a case of shared characteristics that are *analogous* (not inherited from a common ancestor but invented independently – convergently – by each species). We have already flirted with this idea when looking at swallows and swifts, but it may be easiest to grasp with the more blatant example of bird wings and bee wings. Did both these flying animals inherit wings from their 555-million-year-old common ancestor? Of course not! Swift wings and bee wings are at best superficially similar, being flat, flapping appendages that generate lift, but a bee's wing contains no scapula, humerus, radius, ulna – no bones at all.

Another way to grasp the non-homology of bird and bee wings comes from the tree. If bird and bee wings really were inherited from a winged common ancestor, then the wing must have existed in an unbroken chain throughout both of their evolutionary histories. Their distant relationship would then require the wing to have been lost over and over again in all of their wingless relatives – crustaceans, spiders, millipedes, nematode worms, priapulid worms and annelids on the bee branch; dinosaurs, crocodiles, lizards, mammals, amphibians, coelacanths and goldfish on the bird branch. In terms of parsimony, this model – the model that says bird and bee wings are homologous – has a single instance of wing invention (in the ancestor) but a great many instances of loss. The alternative model – that tells us bird and bee wings are not homologs – simply requires that wings were invented twice. This alternative model is much more parsimonious and hence the one we believe.

These examples show how we can use the parsimony principle and our knowledge of an evolutionary tree to work out where (perhaps I mean when!) on the tree characters evolved and, by extension, whether two characters are homologous. In our examples of limbs and wings we knew the tree that related the species, but perhaps the most powerful use of the parsimony principle is when we don't know the tree, when the relationships between species is the very thing that we are trying to discover.

5

Are We Still Fishes?

THOUGH IT OCCASIONALLY feels like it, it is rarely true that getting the tree of life wrong leads to disaster, rarer still that it carries the risk of death; it is rarely but not never true, as the story of the British explorer Apsley Cherry-Garrard proves. Cherry-Garrard was a twenty-four-year-old minor British aristocrat when (with a large donation oiling the wheels) he managed to be included, as an assistant zoologist, in Robert Falcon Scott's second (fatal) expedition to the South Pole. His most notable contribution to the expedition (apart from simply surviving it) was a journey in the most unbearably cold month of the Antarctic winter to collect the eggs of the emperor penguin. He tells the story of his part in the expedition in his marvellously stiff-upper-lipped memoir *The Worst Journey in the World*, a title it is hard to argue with.[1] His nightmare took him and two companions from the expedition base at one end of Ross Island to Cape Crozier, sixty miles away at the far end of the island. It was June 1911, the southern hemisphere's mid-winter, and temperatures ranged from -40 °C to -60 °C; the hardship of the journey in the clothes and equipment of the time – tweed and oilskins and hobnail boots – is hard to imagine. Cherry-Garrard never fully recovered from his Antarctic experiences, which were compounded by his service in the First World War. The image in his book that has stuck most in my mind is of the dampness of the clothing that would just about melt during a night in an inadequate sleeping bag, only to freeze solid within moments of leaving the tent. This was more than an inconvenience; it meant the men had to bend forward in sledge-pulling position immediately on emerging from the tent or else be frozen uncomfortably upright for the rest of the day (not that there was any daylight). All three men survived the nineteen-day journey to the penguin colony and back again; they brought with them three precious penguin eggs, which were eventually delivered intact to the Natural History Museum in London.

The whole terrible exercise was predicated on the mistaken idea that penguins are the flightless missing link between reptiles and birds. This error about the place of penguins on the bird tree of life was compounded by the mistaken idea (one coming from Ernst Haeckel no less) that a study of penguin eggs would reveal a lizard-like embryonic stage in penguin development. Penguins, it has since been shown, are not distant relatives of all the flying birds halfway between earth-bound lizards and the evolution of flight; their true place on the tree of life is as a twig nestled right within the main bird branch (closely related to those very competent fliers the albatrosses and petrels). Penguins, about 60 million years ago, abandoned flying through the air to fly through the water instead.

Knowing (or not) the true shape of the tree of life rarely has such an impact, but, as we are discovering, knowledge of relationships is crucial for reconstructing the events of evolution. The broad outline of how we might ascertain relatedness will be familiar from our example of limb evolution for understanding homology. Returning to our simple example of the fish, bird and human, we can now ask how we might use the presence or absence of legs to decide how these three are most likely to be related to one another. Just as we did before, we are going to use parsimony to find the most plausible of three different trees.

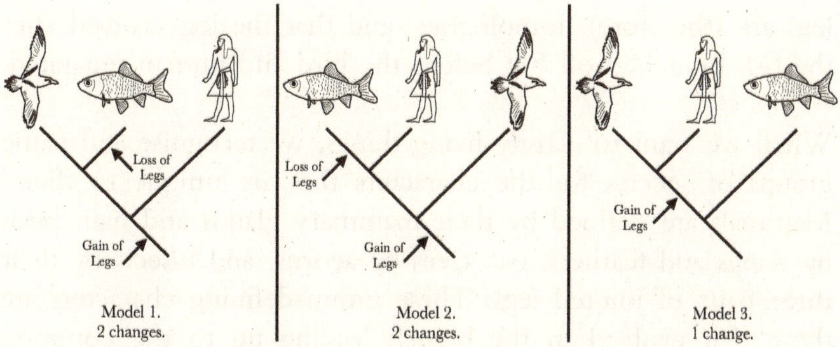

FIGURE 7: Three models of how bird, fish and human are related to one another (with different implications for the evolution of legs). Model 3 is the most parsimonious and most likely.

In the same way as before, we can count the number of times characters (legs in our example) are gained or lost on each of our three different trees of relationships. Our three models are trees in which: (i) the bird is closest to the fish, and the human is more distantly related; (ii) the human is closest to the fish, and the bird is more distantly related; and (iii) the human is closest to the bird, and the fish is more distantly related. On the first two trees we have to assume either that limbs evolved twice – once in the bird and again in the human – or that limbs evolved once in the common ancestor of all three species, but were then lost in the branch leading to the fish. That means that in both the first and the second trees we have to assume that two evolutionary events occurred. By contrast, in the third tree we can end up with both birds and humans possessing a limb (and fish not) by assuming only a single evolutionary event in which limbs evolved on the branch leading to the common ancestor of the human and bird. Tree three is therefore the most parsimonious and our best guess at how these three species are related to one another. We now know several intimately related things: that bird and human are each other's closest relatives; that legs evolved once (not twice convergently); that bird and human legs are (therefore) homologous; and that the leg evolved after the fish branched off but before the bird and human separated.

When we want to classify living things, we recognise and name groups of species for the characters that are unique to them. Mammals are defined by their mammary glands and hair, birds by wings and feathers, oak trees by acorns, and insects by their three pairs of jointed legs. These group-defining characters are those that evolved in the branch leading up to the common ancestor of that group and are therefore also homologous characters that are unique to the group.

The widespread use of homologous characters to define branches on the tree of life is largely attributed to the German biologist Willi Hennig's hugely influential 1966 book *Phylogenetic Systematics*.[2] Hennig was born in Germany in 1913 near Dresden, in circumstances that make it extraordinary that he became a world-famous academic. His family was dirt-poor – his father a railway worker, his mother the illegitimate daughter of a servant (and deeply ashamed of the fact) was 'moody and socially awkward'.[3] The family lived in a tiny house next to the railway, and Willi and his two brothers went to the local school. That Willi made it to secondary school and ultimately to university seems to have been down to a series of minor miracles, combined with his own intelligence and drive; he somehow overcame hurdles such as his own extreme shyness, poverty and the fact that his primary school (now named after him) didn't teach the subjects required to enter the secondary school.

School and university studies led Hennig to become an entomologist (an expert on insects), his specialism being Diptera (the 'two-winged' flies, which include houseflies, mosquitoes and the laboratory favourite, the fruit fly *Drosophila melanogaster*), and he eventually got a job in the Berlin entomological museum. He is rightly famous for his insights into evolution, but his life would be considered eventful and impressive without this achievement.[4] Hennig was an anti-Nazi who fought with the Germans in the Second World War in Poland, France, Russia and Denmark. In

1942 he was seriously wounded by shrapnel on the Eastern Front, and then, in May 1945, in the final months of the war, he was captured by the British on the Gulf of Trieste and held as a prisoner of war until the autumn. Hennig spent these months helping the British deal with malaria using his knowledge of mosquitoes. After the war, and despite his criticisms of its communist government, Hennig was able to work in East Berlin, undertaking a daily round trip of three hours through the iron curtain from his home in the West, until prevented in 1961 by the Berlin Wall.[5]

It was during his time as a British prisoner of war that Hennig began to develop his great contribution to evolution, which insisted on the logic that all branches on the tree of life should be identified (and ideally named) only on the basis of homologous characters. Taking again our example of the limbs shared by mammals, reptiles (including birds) and amphibians, the many detailed similarities of limbs right across this group of four-legged vertebrates (tetrapods) have allowed us to conclude that limbs are homologous in all these animals. For Hennig, this kind of homolog is gold dust because it is an evolutionary novelty.* Because all tetrapods – and *only* tetrapods – have such limbs, this character defines 'tetrapods' as a single group within our classification and as a single branch on our tree. Unique, shared characters are the raw material we need to define each branch on a tree and ultimately build the entire tree of life.

The great German-American evolutionary biologist Ernst Mayr called Hennig and his followers the 'cladists' after the 'clades' of the tree that were defined in this way (*kládos* means 'branch'). Mayr and many others preferred to classify species with a much stronger emphasis on how similar they are. How are these two ways of classifying different? Well, for the most part, they aren't – all species on a given branch are indeed usually more similar

* Hennig used the cumbersome name of 'synapomorphy' for this type of character, coming from the Greek *syn* ('together'), *apo* ('away from') and *morph* ('form'). An 'apomorph' is a new or altered form. And 'syn' tells us it is shared by a group of species.

to each other than they are to other species. Occasionally, however, a classification based purely on similarity can turn out to be very different from a cladistic one.

These two methods of classifying disagree when there is one part of a branch that has evolved to be very different from the rest. The fishes are one of the most famous examples. We all know what characters are typical of a fish – a slender body, gills, fins, tail, scales etc. – and we can probably agree on what species of animals are fishes. These would include trout and goldfish and also, from the description above, the coelacanths we met earlier. The problem with this definition, though, appears when we decide that these characters should be used to define a branch of our tree (and to name it 'fish').

If we start 460 million years ago with the ancestor of all the fishes and follow this branch of the animal tree upwards, we are going to discover that one of the branches that comes from this indisputably fishy ancestor is not (by any common-sense definition) a fish. One of the branches – one that broke off from the rest around 410 million years ago – leads in fact to the tetrapods. So, if we want to define and name the single group of animals that arose from the ancestor of trout and goldfish and coelacanths, we are faced with a difficult choice. What Mayr would do is simply exclude from the fish group all the descendants of the fish ancestor that have evolved legs. What Hennig would do, in contrast, is accept that there is a perfectly respectable group called fishes, but that one offshoot of this branch leads to species that have lost many of the typical characteristics of a fish (and which have evolved to become the tetrapods). According to the logic of Hennig, we humans really are fishes. Hennig, who respectfully disagreed with Mayr ('His indisputably great achievements lie in another field'),[6] has emphatically won this argument. All our classifications of life are now Hennigian (or cladistic), based around these branches, each one defined by its set of unique characters. And it is universally agreed that all groups in our classifications (e.g. fishes) must not exclude any offshoot of that group (e.g. tetrapods) simply because they have lost or altered some of the typical characteristics of that group. The idea of birds as feathered dinosaurs (and so a branch within the reptiles, rather than something quite separate as they were once considered to be) encompasses the same idea.

Tree building, explained in terms of parsimony and the sharing of unique characters, seems like it should be a relatively straightforward project, and yet we have not stopped arguing over the form of the tree of life since Haeckel's first effort. The next chapter will begin to explore these disagreements, and I hope to show how they can tell us something about the odd ways in which evolution occasionally works. We are also going to move beyond morphological characteristics such as legs and wings to think about what other evidence we might use to work out the tree of life. We are going to see that the tree of life is actually written most clearly in our genes, and that our genomes contain the potential to supply us with vast datasets of billions of characters in the form of our DNA.

6

Some Awfully Big Numbers

A PRIAPULID WORM IS not the prettiest of animals. The scientific name of the most common species, one found burrowing in the seabed around the British Isles, is *Priapulus caudatus*, and it is fairly typical of its bough on the tree of life. It looks like a longish, wrinkly, slightly deflated water balloon (which, in truth, is more or less what it is), covered in a leathery, spiky skin. At its front end there is a spine-covered bulb that ends in a hungry mouth, eager to swallow a passing lugworm; at its back end there is a clump of feathery gills. The whole animal is an unappealing off-white colour. Its common name is the penis worm, a name echoed in the only slightly more decorous Latin moniker of Priapulida given by Linnaeus, which also acknowledges its indisputably phallic form. *Priapulus caudatus* will never win a beauty contest, but its downright ornery form is a clear winning formula because the priapulids, like the coelacanths, are little changed from their fossilised ancestors, which lived a little over 500 million years ago.

I want to introduce this phylum of animals not for comedy value, nor even as another example of a living fossil, but because it will help us to grasp one of the biggest problems of working out the structure of the tree of life. Priapulida is one of the very smallest of all the animal phyla. There are only sixteen living species – a trivial sprig of the tree of life when compared to the tens of thousands of annelid worms and molluscs or the millions of species in the arthropods.[1] But working out priapulid evolutionary relationships is still inherently difficult, and just how difficult it is will indicate the unfathomable challenge of scaling up to knowing the structure of the entire tree of life.

In the previous chapters our examples using three species showed how to find the tree that assumes the fewest changes and is therefore most likely to show their true relationships. These three-species examples (using a single character) are clearly as simple as can be, and barely make a dent in resolving the tree of life; but they can be made more realistic by using both more species (let's see how we do with sixteen species of priapulid) and more characters – ideally many more. These two added complications have very different consequences. Adding more characters naturally makes the problem more complex, and the calculations slower. But adding more species introduces complexity of a horrifying scale – it cannot be comprehended even by comparison to the number of grains of sand on all the beaches of a galaxy's worth of planet earths.

To start gently, let's see what happens when we move from three species to four. For three species, we had to count character changes on the three possible trees that could relate them. To add a fourth species, we need to look at these three trees and ask where *on each of them* the fourth species might be placed – in this case we're adding a lizard. If we take one of the three trees (it happens to be the correct one) we will find that there are five different places the lizard could be added (shown in figure 8): (i) next to the human; (ii) next to the bird; (iii) on the branch leading to the human and the bird; (iv) next to the fish; (v) on the branch leading to the bird, human and fish. But the same is true, *mutatis mutandis*, of both of the other three-species trees. What this means is that, while for three species there are three possible trees, for four species there are $3 \times 5 = 15$ possible trees.

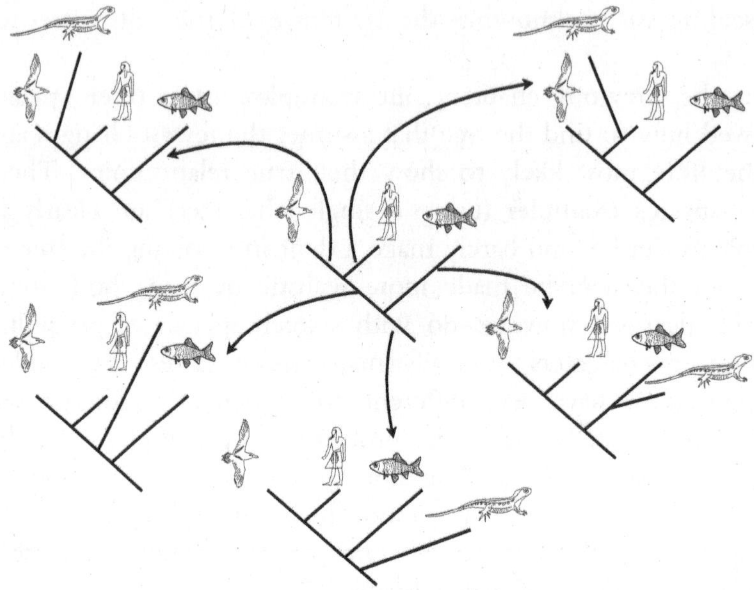

FIGURE 8: A fourth species (lizard) can be added to a tree of three species (bird, fish and human) in five different places.

The number of possible trees grows exponentially and very quickly as we add more species. Adding just one more species results in another even bigger jump. On each of the fifteen four-species trees, a fifth species could be placed in seven different locations, and so the number of different trees that could relate five species is 15 × 7 = 105. For six species, there are 105 × 9 = 945 possible trees. Not only does the number of trees we have to compare increase each time we add a new branch to the tree; the number increases increasingly!

Scaling up the problem from six species to ten, we can calculate that there are roughly 35 million different trees that could connect them, roughly the number of times your heart beats in a year. Jumping forward to fifteen species, the number of trees involved becomes hard to picture – it corresponds to a number (approximately 2×10^{14}, or 2 followed by fourteen zeros) that is slightly greater than the number of millimetres separating the earth from the sun. Can we even do the calculations required on all these different trees? It is just about possible. If we imagine a supercomputer that could do the calculations at the rate of a million trees a second, it would take us a little over six years to find the best amongst all the trees relating fifteen species. The same task with all sixteen species of priapulids, just a single, hideous worm extra, would take 180 years. This is a tiny group of animals; we're obviously already in trouble.

Thinking vastly bigger, the number of atoms in the sun (10^{57}, or 1 followed by fifty-seven zeros) is roughly how many different trees we would have to choose amongst if we were trying to find the optimal tree relating the forty-eight known species of Australian peanut worms (sipunculids). And, taking one final leap, try to conceive of a number that encompasses every single atom in each of the 100 million stars in each of the 200 trillion galaxies in the whole of the visible universe. The vast figure of 10^{80} atoms corresponds to the number of different trees that could relate the sixty-five species of the tiny family of algae called Klebsormidiaceae. I've run out of imaginable big numbers, but the simple truth is that there is no computer we could build that could, in a human lifetime, do the calculations to choose the best tree amongst all the trees that might in theory relate even three dozen species, let alone the millions that make up the tree of life.

The way we deal with this insurmountable obstacle is by abandoning any attempt to climb over it and looking instead for a path around it. This is achieved using a set of shortcuts called heuristics, which are simply tricks that let us skip huge numbers of the possible trees because, for whatever reason, they cannot or are very unlikely to contain the best tree that we are looking for. You could picture this as deciding in advance to ignore any tree with humans and oak trees as closest relatives – clearly it would be a waste of time to do the calculation on any such odd trees of life.

One of the most powerful heuristics is, rather than trying to do the whole difficult problem in one go, simply to build up our tree by adding species one by one. Imagine we are trying to find the best tree that relates six species: we are going to start by considering just three species (it doesn't much matter which three). Our first step is to find the best tree (the one with the fewest changes) that relates these three; because there are only three possible trees to consider, this is easy. The clever part is that from now on we are only going to keep the best three-species tree and we're going ignore the other two less good ones. It might not be immediately obvious, but chucking away these two less good trees means that we have just eliminated from our future calculations two-thirds of all the possible six-species trees.

We now choose at random any one of the remaining species and try to place it on our best three-species tree in all possible places – there are five different places we could add it. Again, we find the best tree amongst these five and discard the other four, thus eliminating four-fifths of the remaining possible trees from future calculations. Adding the fifth species only requires us to consider seven possible places we could add it, and the sixth requires an additional nine. If we are a little lucky, we will have found the best tree having looked at just three trees (for three species), plus five trees (from adding the fourth species), plus seven trees (from adding the fifth species), plus nine trees (from adding the sixth species), to give a total of 3 + 5 + 7 + 9 = 24 trees on which we have to do the calculations. Looking instead at *all* the trees that could possibly relate six species would require us to do the calculations for 945 trees. The benefits of doing things this way grow very, very quickly as we add more and more species. For twenty species, rather than having to do our calculation on all of the 8,200,794,532,637,891,559,375 trees that could in theory relate them, we have to look at just 360.*

* This seems too good to be true, and it turns out it is. This method almost always gets us to a pretty good tree, but, because the order in which we add species can affect where we end up, it doesn't guarantee that we find the best tree. Fortunately, we have other heuristics which can help. One is simply to rerun our process but adding species in a different random order. If we do this 100 times, we will almost always get the best tree.

Increasing the number of characters also makes things harder, but fortunately, the increase in complexity is not exponential but linear. Calculating which is the best tree using ten times as many characters only takes ten times as long, not a trillion times. Using lots of characters is important because some of the characters we can detect in various organisms mislead us: snakes are part of the reptiles despite having no legs; swallows and swifts are not close relatives despite looking the same; and so on. What we are banking on is that the characters that mislead us will be randomly scattered across the branches of the tree while the (hopefully plentiful) truthful, homologous characters will be organised such that they reinforce each other because they share a real evolutionary history. The bottom line is that the more characters we can glean, the better the chance we have of getting the tree right. A second reason for needing to consider lots of characters is that each single character will usually only tell you something about one branch of the tree. In the animal tree of life, the character of having a backbone (or not) can only tell you about what species are in the vertebrate branch (and which are not) and is mute about the relationships between vertebrate species and about relationships between species anywhere else on the tree. Having lots of species on a tree therefore absolutely requires us to consider lots of different characters.

Experts on all different kinds of organism, from slime moulds to flies, have spent long careers trying to find new features to distinguish species and to discover how species might be related. In my own field of reconstructing the animal tree of life, the number of morphological characters that can usefully be considered would be in the order of a few hundred. It is painstaking work to recognise and measure a lot of different bits of morphology in closely related species, but as the comparisons become more distant – human with jellyfish, human with oak tree, human with bacteria – it becomes far more difficult. For the most distantly related species, there is really nothing we can compare beyond features of their cells: there is no plant equivalent of a leg or an eye, no animal version of a plant's leaf or a mushroom's gills.

Tree-builders, endlessly inventive in their search for things to compare, have always been on the lookout for any heritable character that might be useful. So, from the late nineteenth century onwards, as a potent source of new characters emerged from the study of biological molecules and genetics, evolutionary biologists – like seagulls behind a fishing trawler – followed close behind.

Fruit flies, like many other insects, have two very different phases in their lives – two entirely different bodies in fact. The first part of their life is spent as a maggot – in the lab they crawl over and eat their way through a gunk made of agar, yeast, cornmeal and a little sugar (or mushed-up bananas and yeast in the early days of fruit fly genetics). When the time for transition from maggot-hood to adult fly arrives, the first thing they do is climb up the side of the tube that is their laboratory home and put into action a pair of large glands located on either side of their mouth. These are the salivary glands, and they secrete large quantities of 'secretory glycoproteins', which act as a glue for attaching the pupa somewhere out of the way where it can metamorphose in peace. Inside the cells of the salivary glands can be found strange, giant chromosomes, which are really just regular-sized chromosomes that have been copied over and over. This seems to be the fruit fly's way of having many copies of its genes, allowing it to make a superabundance of glue. From a geneticist's point of view, the giant chromosomes are a wonderful way of making fly chromosomes easy to see under a standard lab microscope. The giant chromosomes (when dyed appropriately) reveal a barcode of lighter and darker stripes, some thick and some thin, and the patterns of these stripes allow different regions of each chromosome to be recognised.

Alfred Sturtevant was one of the pioneers of fruit fly genetics. He was ridiculously precocious; he became interested in genetics as a schoolboy, drawing up pedigrees of his family and of horses. As an undergraduate, he interpreted his horse pedigrees in terms of the new discoveries in genetics, sent his findings to the great geneticist Thomas Hunt Morgan and won a place in Morgan's famous 'fly room' at Columbia University in New York City.

Sturtevant's work (on the genetics of horse coat colours) was published while he was still an undergraduate. Even more extraordinary was his invention of one of the most important techniques of twentieth-century genetics – a method to discover the order of genes on a chromosome. He first imagined this new method in the course of a conversation with Morgan; he later wrote that he then 'went home, and spent most of the night (to the neglect of my undergraduate homework) in producing the first chromosome map'.[2] He is famous as a geneticist, but he had other interests, in particular being an expert on the much broader diversity of flies and their classification (beyond the lab species *Drosophila melanogaster* and even far beyond the fruit flies).

Sturtevant's specialities of genetics and fruit fly diversity collided in a paper he wrote in 1938 with another famous American geneticist, Theodosius Dobzhansky. These two titans of genetics studied the giant chromosomes of lots of different populations of a species of fruit fly called *Drosophila pseudoobscura* that they had collected right across the west of the United States. They discovered that in some populations, whole regions of the flies' chromosomes had been neatly chopped out and the chunk then reinserted back into the chromosome in the opposite orientation; this is called an inversion and could be seen as a reversal of a section of stripes in the chromosomal barcode. Dobzhansky and Sturtevant realised that they could think of these inversions as characters that, just like any evolving morphological feature, were passed on from parent to offspring, and that they could use them to understand the relationships between the fly populations. The unique value of the information coming from the chromosomes for distinguishing flies that are otherwise identical is clear in their summing up:

> We have had under observation some hundreds of wild strains, from localities scattered from British Columbia to southern Mexico and from the Pacific to Texas. None of these have been distinct enough so that we should feel certain of distinguishing unlabeled cultures from the external appearance of the flies.[3]

In other words, populations of flies that were indistinguishable on the outside could be told apart – classified into related groups – by looking at changes that had happened to their chromosomes.

Dobzhansky and Sturtevant's study of chromosomes required no knowledge of DNA sequences (in 1938, it wasn't even known that genes were made of DNA) and the genetic material in chromosomes is not the only kind of biological molecule that has been a useful character for defining branches on the tree of life. Any biological feature that behaves correctly can be used. What 'correctly' means is first that the features must occasionally change; second that the changes are encoded in DNA (and so get passed on from parent to offspring); and finally that there is some way of detecting the changes – a microscope or some chemical test. Molecules that have been used as characters for building trees include: carbohydrates, proteins, fats and lipids; enzymes, oxygen carriers, hormones and pheromones; parts of the cytoskeleton (the cell skeleton); and chemical components of shells and teeth and carapaces and bones.

Biochemical characteristics have been a useful complement to morphological characters for reconstructing the tree of life but are, if anything, even fewer in number and more hard-won – almost any biochemical component you might want to study requires its own specialised experiment. What we really want is a source of characters that are quick and easy to read from any given species. We want lots of them. And we want them to be comparable right across the tree of life, from the smallest and simplest to the biggest and most sophisticated. For the past several decades, the characters that fit these requirements, and which have proved to be by far the most useful for reconstructing the tree of life, are the biological information molecules: the proteins and DNA found in all of our cells.

7

Such Stuff as Genes Are Made On

HIDDEN INSIDE THE Statue of Liberty, stretching from the soles of her feet to her crown, you can find an accidental representation of a superstar of molecular biology. It is a pair of narrow spiral staircases, one for ascending and one for descending, each curling around the other to occupy the same central column of space. This double helix was the thing I found most remarkable when I climbed up and down these cramped staircases the first time I visited New York. The double helix is the most iconic feature of the DNA molecule, but it is probably the least interesting aspect of DNA for our present purposes. If we are to understand why DNA (and its alter ego in the form of protein molecules) has turned out to be so useful for reconstructing the tree of life, we need to learn one or two other details of how this information molecule does its job.

The first detail is that the individual letters of our DNA code and the amino acid letters strung together in our protein molecules can be thought about in precisely the same way that we have thought about bits of morphology – flytraps, arms and wings, hairiness, backbones and even chromosome inversions. Each molecular letter can be considered equivalent to a morphological character, and they behave in the same way, changing occasionally as time passes, these changes being inherited by their descendants.

The second detail is something I have so far skated around for the sake of simplicity. The morphological characters of the species we have thought about so far have been either present or absent – backbone or no backbone, snap trap or no snap trap, and so on.

In the real world, if these black and white states were all we were allowed to use to reconstruct the tree of life, we would soon run out of characters; almost all of evolution actually happens not in the invention of entirely new characters but in the elaboration and modification of these characters after they have come into existence.

Take, for example, the teeth of vertebrates. Having (or not having) teeth is an excellent way of deciding whether any given species is a vertebrate (or not). But vertebrate teeth have also changed over time and across the branches of the vertebrate tree. Teeth have evolved in many different directions: long and sharp in carnivores, rough-edged and grinding in herbivores, spiky in fish-eaters, huge in elephants, and on and on. The existence of these many different versions of teeth means there is masses of useful information beyond a simple 'have teeth'/'don't have teeth' dichotomy. All of this added subtlety means we can use the character 'tooth' to understand relationships *between* vertebrates – telling a lion from a tiger or a porpoise from a dolphin.

The various different forms (sharp, spiky, grinding) that a single character (tooth) can assume are called character states, and without this more nuanced concept, molecular characters don't make much sense. Almost all analyses of molecular data use character states: we identify a character (e.g. the fifth amino acid in a protein) that is present in all the species (tiger, lion, sheep and goat), and then, for each species, we ask which of the twenty possible different amino acid letters is present at this position in the protein of each of the species: perhaps the fifth amino acid in the tiger and the lion is the amino acid Alanine; and in sheep and goat a Serine. In this case it is not the having or not having of the character (they all have the character) but the different character states that tells us what we want to know.

With these points in mind, we do need to dip a toe into the working of genes – at least the parts that are important for tree building.

Protein and DNA molecules are both polymers, which simply means that they are each composed of a long, unbranched chain that is made up of lots of smaller subunits all in a row – much like jewels strung together to make a necklace. In a protein molecule, the subunits (the rubies, diamonds, amethysts, opals in a necklace) are amino acids – twenty different molecules, each a slight variation on a central constant theme. They all share the chemical formula NH_2-CHR-COOH; the 'R' indicates the only part of the molecule that varies. In Alanine, R represents 'CH_3', for Serine substitute R with 'OH' (you will not be tested on this). Each different R gives that amino acid its unique set of properties: the R part may be big or small; it may enable the amino acid to dissolve in water or to repel water; it may mean the amino acid is more or less acidic; and so on. Just as the number and order of letters in the sentence you are reading determine its meaning, the number and order of the twenty different amino acids that make up a protein determine what the protein does.

The jewels that make up the necklace of DNA, on the other hand, are called nucleotides. These are slightly larger and more complex molecules than amino acids, but a DNA molecule is simpler than a protein molecule overall because there are just four different DNA nucleotide letters: Adenine (A), Cytosine (C), Guanine (G) and Thymine (T). These names stand as a subtle record of the history of their discovery: Adenine from the Greek *aden* meaning 'gland' – first extracted from a pancreatic gland; Cytosine from the Greek *kytos* for 'cells' – it was extracted from the cells of a thymus; Guanine from 'guano' (originally from the Incan word *wanu*) – extracted from bird droppings; and Thymine purified from thymic acid extracted from a thymus gland. The same is true of the amino acids – Serine was first extracted from silk (*sericum* in Latin); Alanine was first synthesised from aldehyde; and so on.

The structures of these two chemically very different polymers are nevertheless intimately connected: the order of the nucleotide letters in a section of DNA (which we call a gene) encodes the order of amino acids in the protein that the gene produces. Almost all the knotty details of how the DNA code gets translated into the correct series of amino acids are of little importance to the story of the tree of life. The exception lies in tiny molecular machines called ribosomes, which are found in the cells of every living organism. The job of a ribosome is to read the DNA code letter by letter and to translate these DNA letters into the correct series of amino acids.*

The DNA can be thought of simply as the inert instructions that tell cells exactly how to make proteins. It is the protein itself that actually does something useful. Depending on the order of its amino acids, the protein may be an enzyme that helps digest food or a globin that carries oxygen to the muscles, or its role may be to switch another gene on (or off) so that a second protein is produced in the cell.

Unknown to Darwin, who was unaware even of the existence of genes, let alone DNA or the double helix, this chain of events is the molecular basis for how evolution works. A change in the DNA sequence of a gene (a mutation) will (usually) result in a change in the amino acid sequence of the protein it codes for, and this changed protein will (usually) produce a change in the phenotype of the organism. The change to the phenotype may be a change for the worse (resulting in an unhealthy organism, few offspring and the eventual demise of this particular mutation), or it may result in the longer neck required by a giraffe to reach the leaves at the top of a tree. This stronger, better-fed giraffe will have extra baby giraffes, resulting in the spread of both the beneficial new mutation and the longer neck it causes.

If you remember one thing from this discourse on how genes work, let it be this: we can analyse DNA and protein characters (the four letters of DNA, the twenty amino acids of proteins) using exactly the same way of thinking that we used for morphological characters. And the really exciting thing? There are really quite a

* For clarity, I am leaving out many details, but the most glaring omission is probably that the ribosome does not really read the DNA letters directly. In order to avoid getting damaged in the busy hubbub of the cell, DNA remains safely in the nucleus, while more expendable RNA molecules are made as perfect copies of the DNA to be read by the ribosome.

lot of these characters in any given species – roughly 3 billion nucleotide letters in your own genome, amongst which there can be found about 20,000 genes, each coding for a protein with its own unique sequence of amino acids, and all these waiting to be read.

The potential for DNA and proteins to record evolution was first pointed out more than sixty-five years ago by Francis Crick, the famous co-discoverer of DNA's double helix. In 1958, well before the sequences needed to do this were available, Crick wrote:

> Biologists should realize that before long we shall have a subject which might be called 'protein taxonomy' . . . It can be argued that these sequences are the most delicate expression possible of the phenotype of an organism and that vast amounts of evolutionary information may be hidden away within them.[1]

Crick's insight was years ahead of its time. The usefulness of biomolecules for building the tree of life would not be realised until techniques for reading the sequence of amino acids in a protein and, later, the nucleotides in DNA were developed to the point where they were easy to use. Extraordinarily, the very different methods to determine the sequence of letters in both types of molecule were developed by the same British biologist, Fred Sanger, and each of these twin achievements would earn him a Nobel Prize.

Today, almost all efforts to work out the structure of the tree of life depend on comparing the sequences of nucleotide letters of DNA and amino acid letters of proteins. There are (at least) three big advantages to using molecular (genotypic) over morphological (phenotypic) characters. The first and most important, as we have seen, is simply a matter of numbers. Now that sequencing whole genomes has become cheap and easy, molecular datasets typically compare perhaps a thousand different genes across the species they are studying, and this translates into hundreds of thousands of individual nucleotides or amino acids. If DNA sequencing is easy today, it was more difficult when I did my doctoral research in the Department of Zoology at Oxford, starting in 1989. I probably sequenced little more than 20,000 DNA letters in three years of lab work. These days, it would take no more than a few hours in the lab to show you, dear reader, how to read the sequence of millions or even billions of nucleotides within the DNA of any organism you might find interesting. Compare this situation to the many years required to become an expert on the morphological characteristics of any small branch of the tree of life.

The second advantage of molecular data is that many genes are found across all of life; we are famously supposed to share half our DNA with a banana – not true, but we do have hundreds of genes in common. An expert on insects and another expert on mammals are really going to struggle to find more than a handful of morphological characters that can be compared between an ant and an anteater, and the more distant the relationship, the fewer comparable morphological characters we will find. In contrast, there are multiple 'universal' genes, found right across the tree of life, involved in the very oldest aspects of being alive: copying DNA, translating DNA into proteins, making energy and other bits of biochemistry essential to all of life. The existence of universal genes means that there are plentiful characters with which I could compare you to a banana.

The third advantage of using molecules is that they are almost entirely immune to the most important kind of convergent evolution. While two species often evolve to look similar when they have similar requirements, they are extremely unlikely to have made the same changes to the same genes to produce those shared characteristics. For example, both crocodiles and lions have evolved big sharp teeth in order to catch prey, but the many different ways there are to make sharp teeth (millions of different ways to alter hundreds of different genes all of which can result in bigger sharper teeth) make it all but impossible that the sharp teeth genes of crocodiles and lions will also happen to look similar.

One of the first attempts to use sequences of genes and proteins to reconstruct (a part of) the tree of life came in 1958 from the inventor of sequencing, Fred Sanger himself. Sanger compared the sequences of amino acids his lab had painstakingly read from the insulin proteins from five mammals – a sheep, a cow, a horse, a pig and a whale. Like the first steps of a toddler, this work was more momentous than it was successful. The first thing Sanger noted was that the insulin proteins from these five animals are almost identical – only three of their fifty-one amino acids vary at all.

Slim pickings if we want to use changes in amino acids to work out relationships. In what is really a paper about insulin sequencing, rather than the tree of life, Sanger came to one evolutionary conclusion: 'It emerges that whale insulin has the same structure as pig insulin.' But this was immediately dismissed: 'it seems likely that the identity of pig and whale insulins should be ascribed to coincidence rather than to any unexpected phylogenic [*sic*] relationship'.[2] What a shame: with the benefit of hindsight, we know the great similarity of whale and pig insulin is not down to mere coincidence; amongst all the mammals, it has been shown that pigs are rather close relatives of the whales and dolphins. Sanger came agonisingly close to discovering something really surprising about the origin of whales.

Molecular trees of life that we would recognise today really got going in the 1970s along with several new developments: computer programmes for automatically building trees using sequences; tree diagrams that are beginning to look like those we make today; and, most importantly, more and more sequences from more and more species – first proteins and later DNA.

My own interest in all of this started as a zoology undergraduate in the late 1980s, and the key event, interested as I was in the arguments over how major animal groups are related, was a scientific paper published in 1988 – a year before I began my doctoral research – by members of Professor Rudolf (Rudy) Raff's lab in Indiana. The first author of the paper was Katherine Field, and the paper came to be known as Field et al.[3]

The Field et al. paper was the first credible attempt to use DNA sequences to understand how certain groups of animals (obviously distinct groups such as vertebrates, molluscs, insects, flatworms, jellyfish etc.) are related to each other. The Field et al. analysis used the sequences of nucleotides – As, Cs, Gs and Ts – in a gene called the 'small subunit ribosomal RNA', or SSU rRNA.* As we will discover in the next chapter, the SSU rRNA gene has played a key role in working out the tree of life since the very first use of molecules.

SSU rRNA is a member of a special cadre of genes whose DNA sequences do not contain the code for a corresponding protein. The function of these ribosomal RNAs is to form the framework of the ribosome itself. The ribosomal RNA molecules act like the chassis of a car onto which a few dozen ribosomal proteins (which are encoded in normal protein-coding genes) are bolted to construct the molecular machine of the ribosome. The sequences of nucleotides in animal ribosomal RNAs were used by Field et al. for a multitude of reasons: they are easy to extract in a pure form from any organism; their sequences are (for unimportant technical reasons) easier to read than those of other genes; they are long enough to provide hundreds or thousands of characters; they evolve slowly, making a comparison between distantly related animals possible; and finally (because they are a component of the universal cellular machine the ribosome), they are universal genes found in all species of life on earth, meaning that it is possible to compare a human SSU rRNA with that of a bacterium. The Field et al. paper was of great importance; it established a new way of working that would dominate the field of animal phylogenetics for the next decade as people like me added SSU rRNA sequences gleaned from more species. But some of their headline results (most controversially they proposed two separate origins of animals) were the result of errors that we will meet later.

The unique qualities of SSU rRNA genes underlie what may be their very biggest contribution of all to reconstructing the

* I am simplifying for clarity: the product of the gene is SSU rRNA; the gene is SSU rDNA. The RNA letters are slightly different as well, Thymine (T) being substituted by Uracil (U).

tree of life. A universal gene, homologous in every single living organism, points to a common ancestor of all of life and one that is very, very old indeed. At the very bottom of the tree of life, there is a species – Haeckel's *Radix communis organismorum* – that lived about 4 billion years ago. This organism is now known as the Last Universal Common Ancestor, or LUCA, and working out what LUCA looked like, what genes it might have had, where it got its energy, what its cell membranes might have been made of, in what sort of environment it lived and what other qualities it might have possessed is of enormous importance to the project of reconstructing the history of life on earth. LUCA is like the narrow part of an egg timer through which everything must pass – it is the pivot between the time before, when chemistry transformed into biochemistry and life emerged from rocks, and the time after, when the processes of evolution spread the astounding diversity of life across the earth. It is to this pivot point that we will travel next.

PART II

How to Travel in Time

8

Meet LUCA, the Last Universal Common Ancestor

LIFE APPEARED INCREDIBLY, almost implausibly quickly after the traumas attending the formation of the earth. The earth first came into existence 4.5 billion years ago, but then, about 4.4 billion years ago, a second planet smashed into the young proto-earth, liquefying both planets to a boiling lava. The debris thrown out from this unimaginably catastrophic event eventually coalesced to form the moon. There was water on the early earth, but at first there were no oceans, the water existing as clouds of steam. As the earth cooled, clouds condensed to fall as rain forming the oceans, and the conditions for the first steps in the origin of life were met. A less pleasant cradle for life is hard to imagine but the tree of life is thought to have sprouted and gained a precarious root-hold in the depths of these oceans as long as 4 billion years ago.

The very oldest preserved structures that are generally agreed to have been made by living organisms (products of biology rather than of chemistry or geology) can be found in Australian and South African rocks that are roughly 3.4 billion years old.[1] Some of these fossils, called stromatolites, are large, visible with the naked eye. They can be seen as many thin layers of rock, like the leaves of a *mille-feuille* pastry. Stromatolites are thought to have been formed in a shallow sea, each layer a fossilised film of photosynthetic bacteria encrusted with sediment. There are also microfossils – minuscule shapes in the rock made by individual fossilised bacteria. But LUCA must be older than these by as much as half a billion years, because traces of the *processes* of life, if not fossils of life itself, can be detected in the rare rocks of this great age.[2] These traces (or 'biomarkers'), found in extraordinarily rare and ancient zircon crystals from Canada and Greenland, survive in the form of unusual carbon isotopes that can be best explained as a by-product of life. This evidence is like the whiff of perfume from someone who has recently left the room, but we have no more direct evidence of the very oldest life on earth. No way of measuring it, knowing its shape, how it reproduced, how it fed, of seeing inside its cells, extracting its DNA, examining its proteins or probing its biochemistry. To understand this foundational stage in the story of life on earth – to travel back in time to meet LUCA – our only recourse is to extrapolate backwards from its living descendants.

One extraordinary truth revealed by Darwin's theory of evolution is that relationships between species imply the existence of ancestors. Closely related species have recent common ancestors as siblings share parents; more distantly related species have older ancestors as cousins share grandparents. Darwin persuaded most Victorians that this was true for apes and monkeys and for other, stranger animals – worms and flies and jellyfish. But he could see that it must also be true, without exception, for all living beings – fungi, plants, amoebae and on and on right down to the tiniest and oddest of the bacteria – 'that probably all the organic beings which have ever lived on this earth have descended from some one primordial form, into which life was first breathed'.[3]

Ancient though it is, LUCA was not the *first* living organism; it was the result of hundreds of millions of years of evolution and had billions of generations of predecessors. But, out of all these early experiments, it is only LUCA's branch that has survived to bear descendants that are alive today. Every other branch alive during its time has gone extinct. LUCA represents the last moment of this ancestral species before it divided into two distinct daughter lineages which would go on to multiply, evolving to produce all life on the planet today. It is LUCA's two daughters whose identity we now need to explore, because these will be our path to knowing something about LUCA.[4]

Some aspects of LUCA are fairly easy to determine (needing no knowledge of LUCA's daughters), because there are some biological processes that are conserved across the whole of life. The small subunit ribosomal RNA (SSU rRNA) gene, which Field et al. used to study the relationships of animals, exists in much the same form (roughly the same DNA letters in roughly the same order) in every cell ever studied. This universality tells us something important; that LUCA must not only have had a ribosome but that its ribosome was built in much the same way as all of the modern ribosomes which are descended from it. LUCA's ribosome was constructed using the same SSU rRNA and the same set of ribosomal proteins (because these are similarly universal). LUCA must have had a ribosome, but we can take this inference a little further to see that the job of a ribosome – to translate the message coded in the DNA of a gene into the correct series of amino acids of its protein counterpart – must also have been part of LUCA's biology.

Perhaps the most improbable coincidence of all is the universality of the genetic code itself. DNA, you will remember, is a long, linear molecular code written in a four-letter alphabet: A, C, G, T. The information stored in DNA is read (by the ribosome) in groups of three DNA letters (AAA, ACG, TGA and so on – there are sixty-four unique 'triplets' in all), and each of these different triplets codes for one of the twenty different amino acids.* In our cells, AAA codes for the amino acid Lysine; CCA codes for the amino acid Proline; and so on. As we have read genetic codes from across the tree of life, from animals, plants and yeasts to every variety of bacteria, we have found, with a few *very* subtle exceptions, that this genetic code is entirely universal, unchanged through billions of years of evolution and across countless branches of the tree of life.

* Sixty-four triplets code for just twenty amino acids because sometimes more than one triplet codes for the same amino acid. For example, the triplets TCA, TCC, TCG and TCT all code for the amino acid Serine.

That these extraordinarily improbable things are shared by all of life – the exact shape, composition and function of the ribosome; the highly conserved sequence of nucleotides in the SSU rRNA genes; the exact same genetic code – together constitutes the most persuasive proof possible that everything alive today arose from one single origin. The sixty-four triplets of DNA letters *could* have been mapped to the twenty amino acids in a mind-bogglingly large number of ways (10^{84} alternative ways to be fairly precise). It is simply unthinkable that bacteria and seaweeds and aardvarks have, by some extraordinary fluke, independently chosen the same genetic code from all these options. All species on earth today inherited their genetic code from a single common ancestor. There is a single tree of life.

The universality of the genetic code tells us that it was used by LUCA. Some other characteristics of species alive today may be more or less common across the branches of the tree but are not universal and so not inevitably derived from LUCA. To see whether these characters with their patchy distribution can be traced all the way back to LUCA, we have to work a little harder; and a previous chapter showed us how.

We can use our parsimony method to work out the characteristics of the ancestor of any two species (any two branches on the tree) by discovering their shared homologous characters. The method is straightforward – we have used it to discover that the common ancestor of birds and humans had legs (but the common ancestor of birds, humans and fish did not) – but this method only works when we know which two groups of species to compare. We cannot simply compare any two species, because (however different they may appear) they *might* have an ancestor more recent than LUCA. To reconstruct the characteristics of LUCA, we need to find living organisms that lie on the two most distantly related branches on the whole tree of life – the branches that separated immediately after LUCA, about 4 billion years ago.

Since at least the 1960s, textbooks have told us that the most fundamental division of life can be found lying at the junction between two very distinct groups: the prokaryotes and the eukaryotes. Their names reveal the greatest distinction between them: the *karyon* is the cell's nucleus (the word means 'kernel' as in a nut), and prokaryote cells ('before the *karyon*') don't store their DNA in a nucleus – they don't *have* a nucleus – while eukaryotes ('with a *karyon*') do. Almost all the millions of species of prokaryotes are, rightly or wrongly, desperately obscure. The islands in this sea of obscurity tend not to be famous but infamous, known for causing tuberculosis, whooping cough, *Chlamydia*, syphilis, gonorrhoea, typhus, botulism, *Salmonella*, anthrax, Lyme disease, *Shigella*, Legionnaire's disease, bubonic plague, leprosy and so on.

Some eukaryotes, at least, are more familiar; they include, of course, the large and complex multi-celled animals, fungi and plants that are commonly thought of as 'living things'. But the eukaryote group also contains many single-celled (and hence microscopic) species that, until the nineteenth century, had generally been lumped in with bacteria. These single-celled life forms had been given various different names – 'protozoans', 'protists', 'schizophytes'. Linnaeus himself had been spectacularly unambitious with regards to the classification of what is really an incredible diversity of single-celled life, both prokaryote and eukaryote, lumping anything microscopic into a single species (!) which he called – presumably with his tongue in his cheek – *Chaos infusoria*. The microscopic eukaryotes are actually extremely diverse – some are more different from one another than humans are from seaweeds – but, like the prokaryotes, most of the halfway familiar single-celled eukaryotes are known thanks to the diseases they cause: *Plasmodium* (malaria); *Trichomonas* (trichomoniasis); and *Trypanosoma* (sleeping sickness and Chagas disease).

Beyond the possession (or not) of a nucleus, the differences between prokaryotes and eukaryotes are manifold, but they can be boiled down to a simple dichotomy: eukaryotes are complex and prokaryotes – almost entirely defined by the eukaryotic characteristics that they lack – are simple. Prokaryotes lack not just the nucleus but also mitochondria, internal cell membranes, mitosis (for cell division) and linear chromosomes. Prokaryote cells are also much smaller and have far less DNA, smaller ribosomes, fewer genes and so on.

The names of these two independent branches 'prokaryotes' and 'eukaryotes' contain a contradiction.[5] From our modern, cladistic perspective (which, as Hennig taught us, requires names to correspond to entire branches), these two named groups should leave nothing and no one out; the eukaryotic branch should contain its last common ancestor and all of this ancestor's descendants, and the prokaryotic branch should contain its own, independent last common ancestor, as well as all of its descendants. The contradiction arises because the name 'prokaryote' ('before the nucleus') suggests that the tiny, simple prokaryotes are primitive and must have existed early on, and that at some later date the larger, complex eukaryotes evolved. This implies that the younger eukaryotic branch is an offshoot of the older prokaryotic branch. But eukaryotes cannot both emerge from within the prokaryotes and have an independent origin.

Encompassed within this confusion we seem to have two possible trees of life relating prokaryotes and eukaryotes – two trees that would imply very different things for our task of working out the characteristics of LUCA. If the very oldest split in the tree of life is between prokaryotes and eukaryotes, then LUCA would be their last common ancestor and any characters they share will have been possessed by LUCA, too.

If, on the other hand, eukaryotes are just a branch that sprouted from somewhere within the prokaryotes, then we cannot use a prokaryote/eukaryote comparison to discover the characteristics of LUCA. In this second case, the eukaryotes are a recently evolved distraction, and we would need, instead, to delve deep into the tree of the prokaryotes (which would in fact constitute a clade encompassing all life on earth) to discover which two prokaryotic branches are most distantly related – for these would be the two branches that had emerged from LUCA.

It turns out that the relationships between the eukaryotes and different lineages of prokaryotes are even more complicated and surprising than either of these two trees suggests. The story of the unmasking of LUCA – and of the big surprise that lies at the origins of the eukaryotic branch – was begun in the nineteenth century, and its final paragraphs remain to be written. For now, though, we are going to parse the lines of its most important subplot, taken from the first paper to use a universal gene to study the deepest branches of the tree of life.

Sequencing a gene today (or an entire genome for that matter) is a doddle. First catch your organism; grind its cells up; add a few chemicals to remove everything but the DNA; and pop the tube of DNA in the post to be processed by a robot. If today's experiment were a jet plane, the experiments to sequence genes that I did in the early 1990s would be the Wright brothers' *Flyer*. But the work done in the 1970s would be more like crossing America in a covered waggon. These pioneering experiments were slow, exhausting and involved a lot of tedium interspersed with moments of real peril. Some of the very first DNA and RNA sequences (of the universal SSU rRNA gene) were worked out in the lab of the brilliant and (reading between the lines of his obituaries) curmudgeonly American, Carl Woese.

The arcane complexities of the experiments are worth skimming over if only because they reveal the terrors of life in the Woese lab of the 1970s. The ultimate goal was to know the sequence of nucleotides (A, C, G, T) in short snippets of the SSU rRNA genes from various species of prokaryotes and eukaryotes. Step number one was to feed the cells with a food source in which the normal phosphorus atoms (^{31}P) had been substituted by a very slightly heavier – and very radioactive – phosphorus isotope (^{32}P). Phosphorus is a key atom in both DNA and RNA, and, as the bacteria grew and multiplied, the radioactive atoms got incorporated into these molecules, allowing the otherwise invisible molecules to be tracked. A young student in Woese's lab, Kenneth Luehrsen, described how the radiation-detecting Geiger counter 'was always screaming a path to wherever you had been'.[6]

The rest of the experiment involved separating the radioactive SSU rRNA from all the other molecules of the cell; cutting it into smaller pieces using enzymes; separating out these many fragments from one another; before, finally, reading the short sequence of ACGT nucleotides each fragment contained. The method for separating the fragments added considerably to the radioactive horror of the experiment.[7] It was done using an early version of a process called electrophoresis. A droplet of the sample of SSU rRNA fragments would be pipetted onto one end of a sheet of moist cellulose paper and an electric current applied. The negatively charged DNA would now flow through the paper attracted towards the positive electrode, but the trick is that the rate at which the fragments move depends on how big they are. Small fragments get much less tangled in the cellulose fibres and move quickly, while larger fragments move much more slowly. The happy result is that the mixed-up SSU rRNA fragments become beautifully and precisely ordered according to their size, allowing each to be fished out in a pure form. It's like finding the Olympic athletes amongst a crowd of people by getting everyone to run a race.

The electrophoresis was done in a liquid-filled 100-litre tank and used an electrical potential of as much as 8,000 volts. The high voltage put the equipment in danger of disastrously melting, so Woese's lab added gallons of a convenient liquid coolant to the tank: refined kerosene, or paint stripper. 'The analysis of RNA sequences in these ways probably could not be conducted today because of safety regulations,' wrote Woese's collaborator, Norman Pace, with wonderful understatement.[8]

Years of work allowed Woese and his colleagues to compare the SSU rRNAs from a diverse collection of both eukaryotes and prokaryotes. Their most important discovery only emerged because they included in their study several species of a particularly bizarre type of prokaryote called a methanogen, which produces the energy it needs to survive not by 'burning' carbohydrates in oxygen (indeed, oxygen is lethal to them) but by combining carbon dioxide and hydrogen to make methane. Methanogens are sometimes referred to as extremophiles because of their strange habitats – very hot deep-sea hydrothermal vents, hyper-salty and hyper-alkaline soda lakes, buried under 3 kilometres of ice in Greenland and, more prosaically, the digestive tracts of many animals. At the time of Woese's paper, methanogens and their relatives were little known, their unusual growth conditions making them difficult to culture in the lab.

The first surprise from their comparisons of SSU rRNAs was that there were some fragments of the SSU rRNA molecule that were identical in every single species they looked at.[9] This was an astonishing discovery – some parts of a universal gene are themselves truly universal, present in life forms as distantly related to one another as humans and the bacteria in our guts. Parts of the SSU rRNA have been faultlessly passed down from generation to generation and across every species that has ever existed for more than 4 billion years. This must be telling us something important: that these exact sequences of nucleotides are so fundamental to the function of the SSU rRNA molecule that any errors in copying it have resulted in the instant death of its unfortunate bearer – no variations have ever been tolerated.

The second and even more notable discovery – a finding which many people, including, allegedly, Woese himself, felt merited a Nobel Prize – concerns the root of the tree of life. The groundbreaking result was shown not in the form of an easy to interpret tree but camouflaged in the entries of the paper's Table 1 – a dry list of numbers which record, for each pair of species, how similar their SSU rRNAs are. Specifically, Table 1 gives a measure of how many of the short sections of SSU rRNAs were shared between each pair of species. The first thing that is immediately and reassuringly obvious is that the (regular) bacteria are all much more similar to each other than they are to eukaryotes, and that the eukaryotes in turn are all much more similar to each other than they are to bacteria – so far pretty unsurprising.

The astonishing discovery comes from the *other* group of prokaryotes – the methanogens. These oddballs are again all similar to one another and very different from the eukaryotes. The surprise, and it was a true bombshell, was that the methanogen bacteria and the 'regular' bacteria did not form a single group together. Woese's data showed, instead, that methanogen bacteria are *at least as different* from the regular bacteria as they are from eukaryotes. Two similarly simple organisms – regular bacteria and methanogens – both lacking a nucleus, mitochondria, chromosomes, membranes and so on, were shown, with this rather clunky experiment (and in this unremarkable table), to be as distantly related to each other as either of them is to the eukaryotes.

Woese had shown that life is divided not into two domains – prokaryote and eukaryote – but into three equally and utterly distinct branches of the tree of life: regular bacteria, eukaryotes and methanogens. He had uncovered the existence of an entirely unanticipated new domain of life, a branch of the tree of life many times older, larger and in every way more significant than the discovery of a new species, genus, class or even kingdom. Many other species have since been added to the methanogen domain, which Woese and his colleagues named Archaea – the 'ancients' – the remaining prokaryotes being Eubacteria.

This division into three – Eubacteria, Archaea and Eukaryota – doesn't sit quite right with the idea we have of a tree that at each branching point divides into two. Woese divided life into three but went no further; he could not immediately tell how they were linked to one another, which of the three possible trees that could relate them was correct. Is the deepest split between Eubacteria plus Archaea on one side, and Eukaryota on the other – in essence nucleus versus no nucleus? Or is the familiar Eubacteria closer to Eukaryota, and distant relatives of the weird Archaea – the possession of a nucleus being a red herring? Or finally, do the branches emerging from LUCA lead one to Eubacteria and the other to Archaea plus Eukaryota? Pinpointing the characteristics of LUCA, remember, requires knowing the very earliest split in the tree of life – we need to know which of these three trees is correct. The problem boils down to deciding which one of the three branches emerges directly from the root of the tree (the other two branches would logically lie together on the other side of this deepest point).

The answer was provided by the pioneering computational biologist Margaret Dayhoff whose lab showed that the root of the tree of life lies between a Eubacteria branch on one side and an Archaea plus Eukaryota branch on the other.[10] The old idea of a straightforward division into simple prokaryotes and nucleus-possessing eukaryotes is wrong.

The story of the relationships between eukaryotes, eubacteria and archaeans is going to get even weirder, but for now we can use this basic understanding of their relationships to travel back a very long way in time, half a billion years earlier than the very oldest fossils known. Our new understanding of this oldest part of the tree of life tells us that any characters shared by Eubacteria (the first branch from LUCA) and *either* Archaea *or* Eukaryota (the second branch) must have come from LUCA. Let's see what we can learn.

We have seen already how the existence of universal ribosomal genes is a clue that tells us that LUCA must also have had the things that go with ribosomes: translation, a triplet genetic code and several other cellular bits and pieces involved in decoding DNA letters and stringing amino acids into proteins. What other universal genes are there that might tell us about the conditions in which LUCA lived, how it got its energy, how it moved? Disappointingly, while the minimum number of genes life can get by with is in the order of 600, there are only about thirty truly universal genes.[11] Even more boringly, almost all of these are involved in the ribosome, reinforcing the idea of just how important this molecular machine is and has always been to life. If we relax a little, however, and require that for a gene to be considered to have come from LUCA, it needn't be universal but could simply exist in at least a few species on both of the branches coming from LUCA, then we discover quite a few more universal-ish genes. With this lower hurdle, the list of LUCA's genes grows to almost 400, and with this list in hand we can begin to get tantalising insights into LUCA's life.

One of the most revealing of LUCA's genes is the one that codes for an exciting-sounding protein called 'reverse gyrase', whose job it is to add extra twists into the DNA's double helix, making it more tightly wound – more compact – than it would otherwise be.[12] This is rather an esoteric job description, but it proves to be a surprising clue to the environment that LUCA lived in. The DNA in most cells we encounter – our own cells for example – has the opposite state: when compared to standard DNA, our DNA double helix is actually slightly untwisted, and this untwisting helps the enzymes that interact with the DNA to get right into the grooves of the double helix to do their job more effectively. The reverse gyrase gene (and the tightly wound DNA it produces) is found scattered across the tree of life, including in species of both Eubacteria and Archaea. The big clue reverse gyrase gives us about the life of LUCA is that almost all the species that still have this gene live in extremely hot environments – from 40 °C to as high as 122 °C (these species are found in hot springs, geysers, etc.). To put these temperatures in a human context, water at 40 °C feels warm to our hands, and 48 °C can (fairly slowly) give us first-degree burns. The tight winding found in the DNA of these heat-loving organisms prevents the two strands of the double helix from being melted apart by the high temperatures. LUCA's reverse gyrase tells us that it very likely lived in a hot environment.

A second clue to the biology of LUCA comes from a set of genes that produce iron-sulphur cluster metalloproteins, which, as the name (hopefully) suggests, are regular proteins that are combined with a core of iron and sulphur atoms. (The combination of a protein with elements such as iron, zinc, etc. is not unusual – the haemoglobin proteins in our own red blood cells have an iron atom at their centres, for example.) The first and most obvious implication is that LUCA must have lived in an environment rich in iron and sulphur. Slightly more tangentially, an interesting aspect of metalloproteins is their extreme sensitivity to even tiny concentrations of oxygen. Oxygen converts iron-sulphur clusters to unstable forms that instantly decompose, with the result that oxygen is essentially lethal.* This tells us that LUCA could not have lived in a world with oxygen. Happily, an anaerobic LUCA is an entirely sensible idea, because LUCA lived about 2 billion years before the evolution of photosynthesis – the source of the oxygen in our atmosphere.

From the discovery that LUCA possessed reverse gyrase and iron-sulphur proteins, we have been able to deduce that LUCA must have lived in very hot water, in an environment that lacked oxygen and where there was a plentiful supply of iron and sulphur. Our current best guess is that LUCA lived at the bottom of an ancient ocean in an undersea hydrothermal vent from which super-heated water came gushing out. The hot water would have supplied diverse minerals (including iron and sulphur) that allowed LUCA to live entirely independent of the sunlight that powers almost all of life today.

Without a real-life time machine, we can only ever hope to have a very partial view of LUCA; we will never know all of the genes it possessed, nor their exact functions, nor how they cooperated to make a complete living organism. This said, several developments are being inventively combined to transport us back: the discovery in the 1970s of modern-day hydrothermal vents; Woese's new tree of life; the massive expansion since the 1980s of the known diversity of Eubacteria and Archaea; and the recent availability of thousands of genomes from all domains of life are all being integrated to take us back to the very early earth. We are becoming better and better at peering through the small porthole of our methodological time machine to get glimpses of our most distant ancestor.

The methods we have used to reconstruct LUCA can be put to work to reconstruct the characteristics of any of the countless common ancestors on the whole tree of life, but very few of these common ancestors generate the interest and scientific effort that LUCA has. One other exception to the general

* It gets worse: oxygen also reacts with iron to produce ferric hydroxide, which forms a solid rather than being dissolved, and this results in the essential iron becoming unavailable to the cells that rely on it.

obscurity of ancestors – another deemed sufficiently important to deserve a name – is an animal called Urbilateria which lived roughly 555 million years ago, in the seas of the Precambrian.[13] Urbilateria, the 'ancient bilaterian', is the ancestor that gave rise to the bilaterally symmetrical animals (with mirror-image left and right sides). Its descendants include almost all animal species alive today.

9

Head-to-Tail Evolution and the First of the Animals

SOME 538 MILLION years ago the very first inkling of the most spectacular flowering of animal diversity nudges its way into the fossil record. These first fossils are called *Treptichnus pedum*, but, unlike most familiar fossils, *Treptichnus* are the petrified remains not of the animal's body but of the burrows it made while looking for food. These fossilised marks are called trace fossils.

The *Treptichnus* fossils are spectacularly unspectacular, but they are not entirely silent about their maker's identity. The first thing we can notice is that these burrows have a fairly complex shape. They have a series of branches where the animal was probing around, presumably seeking food, and this relatively complex behaviour hints at a reasonably sophisticated brain – smarter than a jellyfish at least. The basic contours of the burrows – roughly tube-shaped but with an obvious front and back, top and bottom – can only have been made by an animal that shares the very basics of its body with you and me: a head and a tail; a belly and a back; and left and right sides. The biggest of all the branches of the animal kingdom (first exemplified in the fossil record in the form of *Treptichnus*) is defined by the possession of these mirror-image left and right sides, and is named Bilateria.

Hot on the heels of *Treptichnus* – in very slightly younger rocks – it is possible to detect an extraordinary and rapid flourishing of this branch of animal life. We witness the sudden appearance, as if from nothing, of a host of familiar-looking animal fossils all united by the possession of bilateral symmetry: snail-like molluscs, crustacean-like arthropods, fish-like chordates and arrow worms and penis worms all but identical to those alive today. The sudden appearance of these bilaterian animals in the fossil record is a mystery that has puzzled zoologists and palaeontologists for 150 years, and our picture of the granddaddy of them all, Urbilateria, remains poorly focussed to this day.

To reconstruct Urbilateria (just as for LUCA or any other ancestor), we need to know which living branches stem from it and what homologous characters these branches share. The first of these twin questions has been answered, and we will meet the two branches in due course, but for now you should know that arthropods such as the fruit fly lie on one branch and vertebrates such as mice and humans lie on the other.

It was the problem of identifying homologous characters in Urbilateria's descendants that long seemed an insurmountable problem. For over a hundred years, the way that a fly, a human, a leech and a nematode worm (and species from all the 25-odd major groups of animals that emerged from Urbilateria) grow from an egg to an adult appeared utterly, irreconcilably different from one another. To consider just one kind of difference: some animal embryos (those of nematode worms for example) faithfully follow a precise set of instructions so that all their growing progeny behave identically – the exact same cells are produced in the exact same place at the exact same time. If they were aiming for a street across town, all would follow the exact same route. Many other species (humans, for example) are much more easy-going – the rules are there but are much more general. All embryos reach the same destination but each one will have taken a slightly different route to get there. The notion that we might be able to discover any common details of the *genes* that govern how embryos develop in these very different animals was all but unthinkable.

In the late 1980s, an extraordinary laboratory-produced phenomenon – a fly with four wings – radically changed this picture. This bizarre animal – as surprising in its way as an eight-legged horse or a three-headed dog – was to become a true darling of biology. Its unveiling was the first baby step in a journey that would lead to the discovery of an unanticipated unity in the mechanisms that build the bodies of every species in the animal kingdom. All animal embryos, however different they appear and however differently their embryos may grow, are following a common set of rules. There is an ancient uniformity that is visible only in the code of their DNA and this, once discovered, could be traced back to their common ancestor: the 555-million-year-old worm named Urbilateria.

For a brief, heart-stopping moment in time in the middle of my undergraduate studies, I thought I had discovered one of these amazing four-winged flies. Before I get to my extraordinary finding (please don't get your hopes up), I want to explain just why I was primed for this discovery. I had been attending a lecture course that explored the exciting research being done into the genetics of how fruit flies grow from eggs to adults. This work had its roots in much earlier research, begun in the 1940s, by the American geneticist Edward ('Ed') Butts Lewis. Lewis carried out the research for his PhD at the California Institute of Technology (Cal Tech) in Pasadena with Alfred Sturtevant, whom we met previously using chromosome rearrangements to work out the relationships between flies.[1] Lewis set out to ask questions about genetics – the mechanics of how genes and chromosomes work – but his open mind led him into the very different field of developmental biology and to the study of how genes tell growing embryos how to form an adult organism.

The fruit fly had originally been chosen as the ideal animal for some of the very first experiments in genetics by Thomas Hunt Morgan (Sturtevant's own PhD supervisor).[2] Flies have a rapid ten-day life cycle (no waiting ages for pea plants to produce peas, as Mendel had done) and are very easy to raise. Fruit fly genetics had been launched in 1910 when Morgan discovered a mutant male with white eyes rather than the usual red.[3]

The existence of any given gene is impossible to detect simply by looking at an individual of a species. Looking at a fruit fly's red eyes, we could make a sensible guess that there must be a gene that makes its eyes red, but how could we prove this? How would we go about studying this hypothetical red-eye gene – finding out which chromosome it is on, what other genes are close to it, how many other genes are necessary to make eyes red? It was only thanks to the fortuitous appearance of a mutation that made the fly's eyes white that it was possible for Morgan to know of the existence of a gene that *normally* makes eyes red. Only with such a visible marker of the gene's existence could he study the way it gets passed on to the next generation, its location on a specific chromosome and so on.

Through decades of painstaking experiments, Lewis discovered mutations that pointed to a group of genes all clustered together on the fruit fly chromosome.[4] Mutating each of these genes in turn transforms one part of the fly's body into a different part normally found elsewhere: an antenna might be transformed into a leg to produce a fly with a leg growing out of its head (this gene is called *Antennapedia*).

The most famous of Lewis's genes is called *Ultrabithorax*, and its effects can be most readily seen in the fly's wings. While almost all insects have four wings, all the true flies (Diptera – including fruit flies, house flies, mosquitoes, midges and crane flies) have just two. The two wings are found on the second of the three leg-bearing segments of the fruit fly's thorax; on the third thoracic segment (where all the other insects have their second pair of wings), flies have tiny club-shaped structures called 'halteres', which whirl around like gyroscopes to stabilise flight. When *Ultrabithorax* is mutated, the extraordinary effect is to transform the tiny halteres of a normal two-winged fly into a second pair of wings. The result is the iconic four-winged monster.

The close clustering of these genes on the chromosome, along with their similar effects, also led Lewis to guess that they had evolved by duplication from one original gene. Lewis's famous (and Nobel Prize-winning) genes are called the Hox genes. A second, rather odd discovery was that the order of the Hox genes on the chromosome (we now know there are eight of them in total) corresponds exactly to the order of the body parts each one controls – the first gene was responsible for the cells at the front of the head and the last for the cells at the tip of the tail.

Our interpretation of the effects of Lewis's mutations – and therefore of the normal job of the fly Hox genes – is that each one of the eight Hox genes *normally* gets switched on only in the cells of its own specific part of the fly embryo. The job of each Hox gene is to tell its own set of cells (head or thorax or abdomen) exactly where they are located so that they know which structures to make (the cells that form an appendage should make an antenna if on the head but a leg if on the thorax for example). When one particular Hox gene is mutated in a fly (in effect switched off), the cells it normally instructs don't get the information that tells them where they are. The result is that they make a body part corresponding to the wrong part of the body.

Learning about Ed Lewis's work, I had become fairly obsessed with Hox genes and, with the naïveté of a twenty-year-old, actually hoped I might find a Hox mutant. The first one I believed I had spotted was in a cobweb outside my student room: a spider that had six legs in place of the usual eight. Was this a previously undiscovered Hox mutant? Saving myself some embarrassment, I quickly worked out that I had simply found an old spider who had lost two legs – it wasn't even the same leg absent on left and right. No real shame there; but I came much closer to disaster soon after when I was sorting through a petri dish of anaesthetised flies.

At that time, I spent most of my spare hours between lectures in the genetics department, where I was tackling a project in the lab of the brilliant Professor David Roberts. Roberts – friendly, patient, encouraging and clog-wearing – had a lab studying fruit fly genetics. The aim of my project was to use genetic crosses to try to discover the location on the fruit fly chromosome (amongst the thousands of other genes) of a gene called *Penguin*. The *Penguin* mutant fly has the single pair of wings typical of true flies, but instead of beautiful, transparent, veined plates, its wings resemble stumpy, wrinkled bags like empty crisp packets. Most of my time in Professor Roberts's lab was spent looking down a microscope sorting through dishes of flies made sleepy with ether to find suitable adults. I needed to mate male flies that had the *Penguin* mutation with female flies that had other mutant genes – *curly* wings, or *ebony* bodies or *cinnabar* eyes – in order to see whether, as the animals mixed up their chromosomes when they mated, my *Penguin* mutation travelled along with one or other of these other known mutations. *Penguin* must be grouped on the chromosome together with its fellow travellers, whichever they might prove to be.

It was while looking down my microscope at the sleepy flies that I came across the real deal – an astonishing four-winged fly. A complicated mutant that had taken Ed Lewis decades to breed had appeared spontaneously in my petri dish – in retrospect an impossible event that would have required multiple incredibly rare mutations all to have occurred in a single fly. I was within moments of rushing in to tell Professor Roberts when I saw the truth. My four-winged fly was even more amazing – it had not six legs but twelve, not one head but two. My miracle was a beast with two backs; it was of course simply two flies, one male, one female, made insensible by the ether while *in flagrante delicto*. As my pounding heart sank, the small consolation was that I had kept this to myself – until now.

Lewis's four-winged fly gave a powerful insight into how genes operate in a fly embryo to build the body of an adult insect, but the Hox genes have turned out to be much more interesting (and Lewis's work has spawned many thousands of papers mentioning these genes). The wider importance of Hox genes only really emerged when the sequence of DNA letters in several of them was first deciphered. This experiment (not easy in the 1980s) showed that Lewis's guess that the Hox genes must be related to one another by gene duplication was correct: while lots of the DNA letters in each of the Hox genes are unique to that gene, *all* of the Hox genes share one very similar stretch of 180 nucleotides (called a homeobox).[5] But it was when scientists (including the supervisor of my doctoral research, Professor Peter Holland, during his own PhD) began to look for evidence of homeoboxes in *other* animals that the real surprises appeared.

The first surprise was the discovery of genes containing the same 180-nucleotide homeobox in other species of animal. Homeoboxes were discovered in the DNA of other fruit flies, and in other insects, as one might expect, but they were also found in distant relatives: an earthworm, a chicken, a mouse,[6] a human, sea urchins, starfish, snails and sea slugs.[7] To understand just how surprising this was we need to step into the shoes (or clogs) of a developmental biologist of the early 1980s. For such a scientist, remember, the respective embryonic development of a fruit fly and a mouse appeared, in almost every way, so irreconcilably different that it was impossible to imagine that the head-to-tail axis of a slug and a mouse might be built using the same genes as the head-to-tail axis of a fly. The coincidences, however, began to pile up. Papers published in the following years showed four further stunning similarities between the embryogenesis of mice and flies. The first was that not only were there multiple homeobox genes in a mouse but the DNA sequences of each mouse Hox gene corresponded almost exactly with one of the Hox genes from a fruit fly. The Hox-1 gene of the mouse is most similar to the *labial* Hox gene in the fly; the mouse Hox-2 gene is very similar to the fly gene *proboscipedia* and so on.* The second coincidence was that the corresponding Hox genes in fly and mouse are also found in the exact same order on the chromosome in both species. Fly *labial* and mouse Hox-1 are first; fly *proboscipedia* and mouse Hox-2 are second; and so on. The third was that the link between the order of the Hox genes on the chromosomes and the parts of the body they affect (known from the fly) is perfectly replicated in a mouse.[8] The final coincidence – by this time perhaps not a surprise – was that mutating

* I am simplifying in (at least) three ways: (1) the original names of the mouse Hox genes were horribly complicated and based on the order in which they were found; (2) the mouse Hox complex itself has been duplicated, twice, so (with some additional simplification) there are four copies of Hox-1, four copies of Hox-2, etc.; (3) the correspondence is not exactly 1:1, but it's close enough.

mouse Hox genes causes the exact same kinds of homeotic transformations (one region of the body adopting the identity of another) that are seen in flies.[9]

The only sensible explanation for all these amazing coincidences – same genes, same order, same function – is that both the genes themselves and the jobs that they do must have been inherited by both mice and flies (and humans, slugs, earthworms and all the other bilaterally symmetrical animals) from their common ancestor, Urbilateria. Urbilateria must have had these eight genes, and these genes must have been arranged in the same order along the chromosome and performed the same roles in the same regions of the body as it still does in its living descendants. The extraordinary conservation of this ancient mechanism for creating the head to tail of an animal's anatomy – barely changed in half a billion years – revealed a wholly unanticipated unity that underlies the way all of Urbilateria's descendants are built.

The discovery of the role of the Hox genes was soon followed by work showing that the genes that make the backs of both flies and vertebrates different from their bellies are also conserved (and so also inherited from Urbilateria). And the involvement of a single gene called *Pax6* in making the extraordinarily different eyes of bilaterally symmetrical animals – the eyes of arthropods with their hundreds of individual lenses; our own 'camera' eyes; the superficially similar camera eyes of octopuses and squids; the 200 reflector telescope-type eyes of scallops; and even the tiny, two-celled eyes of worm larvae – tells us that Urbilateria itself must have had some form of eye made using *Pax6*.[10] The *Pax6* gene is so perfectly conserved between insects and vertebrates that if you take the *Pax6* gene from a frog and move it into the genome of a fly (in such a way that it gets switched on), it will produce perfectly good (fly) eyes – multifaceted, bright-red and in all respects indistinguishable from the real thing.

Lewis's four-winged fly has led to a fuzzy blueprint for how Urbilateria was made – we know about its body axes, its eyes. It may also have had a pumping heart (recognised by another ancient gene named 'tinman' – after the *Wizard of Oz* character of course) and a brain and a nerve cord that were organised in much the same way as ours. But, despite these commonalities, the surviving offspring of Urbilateria are so varied (flies, mice, earthworms, snails and so on) that we are still left arguing about what it really looked like. In particular, we disagree about how large and how complex Urbilateria really was. As we will see, resolving this problem has implications for other questions we might have concerning the history of the first bilaterian animals.

10

How the Insects Got Their Wings

'IN NO OTHER part of the Animal Kingdom is the organisation for flight so perfect, so apt to that end, as in the class of insects. The swallow cannot match the dragon-fly in flight,' remarked Richard Owen.[1] The insect wing really is a marvel of evolution. Its invention turned an ancient insect's flat, two-dimensional world into one with three dimensions – the original bird's-eye view was actually an insect's eye view. Flying is much more efficient than walking, and the insect wing expanded its lucky owner's horizons from a few square yards to many square miles – quite why flight has evolved in so few groups is a bit of a puzzle.

Over hundreds of millions of years and across countless twigs of the insect branch of life, the wing has elaborated on its original form (and on its *raison d'être*), changing and adapting to become much more than an object with a single task. Wings have become coloured and patterned, attracting mates, warning predators and acting as camouflage. Beetles and earwigs have changed their forewings into a tough carapace that covers a pair of delicate, flapping hindwings. The flies, as we have seen, have modified their hindwings into the gyroscopic halteres. Halteres are also found in the males of the impossibly tiny, parasitic strepsipterans ('twisted wings'), but in this case it is the forewings that have been converted (the female strepsipteran, meanwhile, passes her entire life inside the body of her host and has abandoned wings entirely, along with her superfluous antennae, legs and eyes).[2] Ants have wings that fall off when they are no longer needed and many stick insects have taken this even further, never growing wings in the first place (flight being unthinkably undignified for these gangly animals). Locusts use their wings as violins; diving beetles as an aqualung; and some tropical dragonflies as radiators to cool their bodies.[3] Even the way that wings are used for their most essential role of flight varies across the insects, from the chaotic flapping of a butterfly to the whine of the mosquito, the glide of a dragonfly to the reluctant, lumbering whir of a ladybird – horses for courses.

Insect wings are a wonderful, transformative invention, but it is hard to imagine how they first developed. What, if anything, did a wing do before it became useful for flight? What, in other words, did the non-flying ancestor of the insects have in place of wings? To discover this, we need to go back to the time just before wings evolved to meet this wingless insect ancestor just as we have done for LUCA and Urbilateria. To do this, the first thing we need to know (because the tree is key) is how insects, which have wings, are related to the other arthropods, which do not. The answer was one of biggest surprises of animal classification in the 1990s and one of the earliest triumphs of using molecules to make evolutionary trees. The closest relatives of insects, it turns out, are the crustaceans – crabs, lobster, shrimp, water fleas, barnacles, woodlice and countless others, almost all of them aquatic.[4] The surprise comes because, since the time anyone had thought about it, insects had been thought to be close relatives of a rather different group of arthropods – the myriapods (the 'many legs', i.e. centipedes and millipedes). Insects and myriapods were thought to be linked because they share a set of characteristics that are absent in all other arthropods: both have only a single pair of antennae (crustaceans have two pairs); both breathe air through holes in the sides of their body called trachea (crustaceans breathe using gills); both have unbranched legs (the legs of crustaceans typically have side branches coming off the base); and both have organs connecting to the back end of their gut that absorb precious water from their faeces before it is expelled (crustaceans do not).

It was a great surprise, therefore, when insects were shown to be not simply *next* to the crustaceans on the tree of life but a group right *inside* the crustaceans, meaning that, in the same way that birds *are* dinosaurs, and we *are* fish, insects are simply a group of crustaceans that has moved onto the land, losing a few pairs of legs on the way. The characters shared by insects and myriapods can now be seen, if anything, as rather more interesting; they were almost certainly invented independently in each group when they first made their moves from the water to the hot, dry, oxygen-rich environment of the land. This new view of the relationship of insects to the other arthropods tells us that the precursors of insect wings will be found by looking not in the myriapods but in the crustaceans.

We've thought, at least glancingly, about where new parts of organisms come from. Rather than inventing from scratch, evolution tends to upcycle something that already exists – a bird's wing from a foreleg; hair and feathers from scales; the delicate bones of your middle ear from massive parts of a jaw bone (still present in living relatives and fossil ancestors). Can we see this process in the insect wing? Can we discover some body part in their closest relatives – the crustaceans – that is a plausible precursor to a wing? This would be a structure which is absolutely not a wing but is nevertheless homologous to wings (in the same way that a bird's wing is homologous to a dinosaur's foreleg).

The precursor of the insect wing has been puzzling zoologists for at least 200 years. One idea was that the wings of the first flying insect started out as the flared edges of the animal's carapace and that this allowed it to glide – like a flying squirrel. A more likely wing precursor, however, is to be discovered at the base of a crustacean leg. Next time you are lucky enough to eat a lobster, you can inspect its ten walking legs. Right where each leg is attached to the body, you will find a cluster of rather unappetising, brownish-grey, feathery fronds. These are the lobster's gills; they are hard to see from the outside because they are hidden safely up under the lobster's carapace. These gills take various different shapes depending on which leg they are attached to, but each gill essentially resembles the thin, flat blade of a table tennis bat; these are the most likely precursors of the insect wing. Gills and wings are found in roughly the same place on the body of a lobster and an insect (closest to the base of the leg), but there are other complex similarities – the equivalent of the set of bones shared by a human arm and a pigeon wing. Most usefully, many of the same genes are switched on in the cells of gills and wings.[5]

How the transition from gill to wing was made is unclear, but it is tempting to speculate. Many of the first insects lived the early portion of their lives in water as mayflies, mosquitoes and dragonflies still do today. They may well have kept the gills from their aquatic stage when they emerged from the water. The first land-dwelling insects perhaps used these paddles to glide a short distance through the air, to avoid a hungry amphibian. Even poor gliding abilities could have been useful at a pinch, and natural selection would next have caused these paddles to expand; to adopt a broader shape; eventually to move, helping the animal manoeuvre in flight; and finally, to flap, allowing powered flight. The proto-wings of the first insects had begun the transformation which would end in the two pairs of broad, flat, flapping structures that are familiar today.

From these simple but functional beginnings, the story of insect wings is one of endless and beautiful variations on a theme – the reappropriation of this basic anatomical feature, beefing up the muscles, adding colour and changing shape in response to new demands on survival. But what if we are still unsatisfied? What if we want to know where the crustaceans got their gills in the first place? What, in other words, did wings look like before they were even gills?

Outside of the crustacean/insect group, there are just four living branches of the arthropods: the myriapods, which we have met; the chelicerates (which pretty much equates to the eight-legged arachnids); the rather distantly related onychophorans (velvet worms); and the microscopic tardigrades (water bears). Of these, only the chelicerates have branches coming off the base of their limbs, but these are already gills so of little interest. To find an answer to where crustacean gills came from, we need to journey even further back in time. A possible answer lies in a branch of the tree of life that has been extinct for close to half a billion years.

The Cambrian period might have begun quietly with the unshowy arrival of the anonymous worm that made the *Treptichnus* fossil marks, but within a 20-million-year geological blink of an eye rocks from the Middle Cambrian record the arrival of an extraordinary diversity of animal types. No vertebrates yet, but molluscs and annelids, arrow worms, lamp shells and, above all, a host of animals related to the modern-day arthropods – distant relatives, that is, of lobsters and fruit flies.

Most of these Cambrian animals crawled across or through the sediments of the sea floor as they probed for food; others led a quiet life, rooted to one spot, sucking nutrients out of the water. Some were algae-eating herbivores and others were predators, but the Middle Cambrian oceans were ruled by just one family of animals – a group of early arthropods called the dinocaridids, or 'terrible crabs'. These magnificent animals didn't crawl across the seabed but flew through the water, cruising above the sea bottom like a golden eagle searching for a rabbit to grab.

The bizarre story of the discovery of the first of the dinocaridids was scattered, at first, across three very different fossils discovered in Cambrian rocks.[6] The first of the three to be described was given the name *Anomalocaris* ('weird shrimp'). It really does look very much like a shrimp we might toss on a barbeque, although the fossils all seem to be missing their heads. The second fossil is called *Peytoia*, which resembles a flattened pineapple ring; a disc with wrinkles radiating from a central opening to the outer edge. Its simple round form led *Peytoia* to be classified as a kind of jellyfish. The third fossil is called *Laggania*, which, being very poorly preserved, was linked confusingly either to the sea cucumbers (relatives of starfish) or to the segmented annelid worms (lugworms and earthworms).

The first clue to the true relationship of these three fossils came in 1978 when the Cambridge palaeontologist Simon Conway Morris re-examined *Laggania* and discovered a shape at the front end that strongly resembled *Peytoia* (the supposed fossil jellyfish). Conway Morris reinterpreted his *Laggania* as a horrible artefact: the decomposed body, he thought, was actually a sponge, and the ring a jellyfish that had fortuitously died atop the dead sponge shortly before both became entombed in mud and fossilised.

While fossils give unique information for reconstructing evolution, the problems they can produce can be seen (inaccurately quoted here with edits for clarity) in Conway Morris's paper on *Laggania*:

> Walcott [described] four new genera, *Eldonia*, *Mackenzia*, *Louisella*, and *Laggania* all of which he placed in the sea cucumbers. However, *Mackenzia* is now accepted as a jellyfish . . . *Louisella* has been shown to be a priapulid worm. And *Laggania* cannot be interpreted as a holothurian [sea cucumber]. Only *Eldonia* remains as a sea cucumber.[7]

Today, even *Eldonia* seems unlikely to be a sea cucumber.[8] Reconstructing fossil animals, especially those for which we have no modern equivalents, can sometimes be like reading tea leaves.

The truth emerged three years later, when Conway Morris's PhD supervisor Harry Whittington, with another student, Derek Briggs, prepared a new *Laggania* fossil by carefully chipping away some of the covering 'matrix' (the rock that surrounds a fossil).[9] As they did so, they discovered first one and then a second fossil of the headless shrimp *Anomalocaris* that were both clearly attached to the *Laggania* specimen's front end. Between these two fossils, and slightly further back, was a specimen of another 'species' – the 'jellyfish' *Peytoia*. What Whittington and Briggs had found was an animal whose body was one species – *Laggania* – which had a pair of large appendages at the front of its head identical to *Anomalocaris* specimens and, between these, a circular mouth resembling a third fossil species, *Peytoia*. All three fossil species were, in reality, parts of a single animal whose fragile body had fallen apart on death, leaving intact just the tough, easily preserved head appendages and mouth. This animal we now know as *Anomalocaris*, which took the name of the first discovered of the three.

The dinocaridids, of which *Anomalocaris canadensis* was the first described, were huge animals, much bigger than modern arthropods. *Anomalocaris* was the size of a cat, but the group would produce absolute monsters; the largest known, *Aegirocassis benmoulae* was 2 metres long, the size of a mako shark.[10]

Dinocaridids were beautifully built for their lives as active hunters in the Cambrian seas: at the front was the head with a prominent pair of eyes and spiny grasping arms ready to grab food to jam into the '*Peytoia*' mouth; behind the head was a body comprised, as in all other arthropods, of a series of segments, on each of which were found two separate pairs of flap-like appendages, one closer to the animal's back (dorsal) and one further down closer to the belly (ventral). The ventral flap is generally leg-like, but the dorsal flap is more interesting, and here, after our diversion into the trials and tribulations of palaeontology, we return to the origins of the crustacean gills. The series of dorsal flaps formed a continuous structure running right along the length of the body, the back part of one flap overlapping the front part of the flap in the next segment along, just as in a bird's wing the separate feathers overlap to make a continuous surface. The whole structure running from front to back on the left and right sides made a long flexible flap that computer simulations have shown would have functioned as a kind of underwater wing, similar to the body-length fins of a stingray or the fringe round the body of a cuttlefish.[11]

These two pairs of flaps on each segment of *Anomalocaris* – those closer to the belly and the dorsal wing-like flaps – are thought to be the precursors of the lobster's legs and gills. These two are separate structures in the dinocaridids but evolution seems to have moved them together and fused them in the crustaceans.[12] Evolution took the *Anomalocaris* water-wing on a strange journey, spending perhaps 100 million years in the form of a crustacean gill before changing once again to become the air-wing of an insect.

We have seen in these examples how we can use knowledge of the relationship between any two species, hand in hand with an understanding of the characters they share, to discover the characteristics of their thoroughly extinct ancestor. As we have seen in the case of the insect wing, following the fate of a character through time allows us to witness its beginnings, the different uses it has been put to in different branches of the tree, and sometimes its demise or repurposing into something brand new. Reconstructing a *series* of ancestors and following them through the tree of life means we can add the dimension of time to our reconstruction of evolutionary history: if a single ancestor is a photograph, a series of ancestors makes a film. We always have to remember, however, that the accuracy of these snapshots and the correct ordering of the individual frames of the film depend on having an accurate tree of life.

We will come to explore the mistakes we have made, and continue to make, when reconstructing the tree of life, but first I want jump back in time to deal with the unfinished business of the surprising origin of the eukaryotes.

11

The Microscopic Hitch-hikers Responsible for You and Me

With a large dollop of luck alongside a willingness to stare down at the sand while you promenade, you may find, on certain northern European beaches, what appears to be skeins of snotty green algae. This green snot gives away its true identity only if you approach it. Sensing your footsteps, it vanishes, seeming to dissolve into the sand. This is not an alga but an animal called *Symsagittifera roscoffensis*. Its genus name *Symsagittifera* comes from the microscopic arrows (*sagitta* means 'arrow') that can be fired out of its skin cells to discourage predators.[1] The species name honours the seaside town of Roscoff in north-west France on whose sandy beaches it can be found in huge numbers.[2]

Symsagittifera is an oddball of the animal kingdom. It is a tiny animal, at most 5 millimetres long, and it derives some degree of notoriety (in the circles I move in at least) from the extreme simplicity of its little green body. It is remarkable because it lacks most of the basic organs that other animals take for granted: a nerve cord and brain; excretory cells, which do the job of a kidney; an anus at the end of its gut – not that it actually has a gut, but rather a solid mass of cells where a gut ought to be. All rather surprising, but the real USP of *Symsagittifera* is that it has a mouth that doesn't eat. This is an animal that is perfectly able, like a genuine breatharian, to live its life – moving, growing, reproducing – entirely without food.

A clue to how *Symsagittifera* manages this trick can be detected in its common name of mint sauce worm. *Symsagittifera* worms are always found in vast swarms, and viewed *en masse*, they do indeed look like mint sauce, with bright-green swirls, eddies, spirals and ripples made of a mass of tiny green dots, in a kind of abstract pointillism that emerges from these murmurations of worms. Mint sauce worms survive on the nutrition that is supplied by the thousands of algal cells that live inside them. The one function of their otherwise pointless mouth is, soon after the worm hatches from its egg, to swallow some of these algae.[3] Mint sauce worms have been called 'plant animals', and this is a good description – they and their algal guests make up a composite organism, part animal and part alga.

One uncomplicated thing we can say about a tree (I mean a real tree – an oak or an ash) is that the branches split apart as they grow upwards, and, once separated, 'never the twain shall meet'. The permanent separation of branches has also been a useful rule of thumb for the tree of life. This separation is simply the manifestation of a really important biological truth – that, albeit with a host of usually minor exceptions, a member of one species cannot reproduce with a member of another. This reproductive isolation allows us to define what a species is. Knowing whether two individuals can mate to produce offspring is how we decide whether they belong to the same or to different species.

There is a very good reason for species to maintain their distinctness, and this is that mixing genes is almost always a mistake. The genes and chromosomes of every species have evolved to work together perfectly like the cogs, wheels, jewels and springs of a Rolex watch. To try to combine the precisely engineered parts of two different makes of watch would be to build a Frankenwatch that, if it even ticks, is unlikely to tell the correct time more than twice a day. Because cross-breeding generally has very poor outcomes, species have evolved clever ways to avoid reproducing with each other. Reproductive isolation takes many forms: geographic or temporal isolation (not breeding in the same place or at the same time); having unique cues for reproduction (a species-specific song or dance or a pheromone that attracts the opposite sex but which other, even closely related species do not find alluring); mechanical isolation, whereby the male genitals of one species don't fit into the female genitals of the other; and incompatibilities between gametes – the eggs of one animal species rejecting the sperm of another, or the pollen of one plant species failing to grow a pollen tube in the stigma and style of another. Even if two closely related species can overcome all of these barriers, when the resulting hybrids are born, they are likely, like the mule (the offspring of a female horse and a male donkey), to be infertile. Aristotle, not surprisingly, knew about mules, but he is often said to have believed in much more extreme cross-species hybridisation (an idea linked with his phrase *Ex Africa semper aliquid novi* – 'Out of Africa new things always come' – the very phrase that began the paper describing the coelacanth).* According to Aristotle,

> in that country [Africa] animals of diverse species meet, on account of the rainless climate, at the watering-places, and there pair together; and that such pairs will often breed if they be nearly of the same size and have periods of gestation of the same length . . . They say that the Indian dog is a cross between the tiger and the bitch . . . They take the bitch to a lonely spot and tie her up: if the tiger be in an amorous mood he will pair with her; if not he will eat her up, and this casualty is of frequent occurrence.[4]

Aristotle's ideas of the mating of a lion and a leopard, or a dog and a tiger, seem just about plausible (and little more outlandish than cross-breeds such as the rather cute 'golden dox' – offspring of the hopelessly ill-matched dachshund and golden retriever). Other closely related species can hybridise – a zebra with a donkey (zedonky) or a lion with a tiger (liger). The DNA of modern humans, *Homo sapiens*, also contains evidence of past

* The common idea that Aristotle also believed the giraffe was the offspring of a camel and a leopard seems to be a relatively modern confusion of his watering-hole theory with what is really a simple description of a giraffe as being camel-like with leopard-ish spots.

hybridisations with *Homo neanderthalensis* and *Homo denisova*. Glossing over these exceptions, though, it is easy to agree with Aristotle that the more different (and distantly related) two species are, the more surprised we would be to find that they had successfully mated.

With this in mind, it is easy to imagine the chin-scratching scepticism that greeted the proposal, in 1967, by the American scientist Lynn Margulis* of the hybridisation of two species whose last common ancestor was 2.5 billion years old.[5] We struggle to believe in a successful hybridisation between a dog and a tiger, still less between a dog and a lizard, or a dog and a fish; but Margulis claimed to have found evidence for a hybridisation between two species that were roughly five times more distantly related than a dog is to an octopus. Margulis has told how her paper, although on the whole quickly believed once published (in the *Journal of Theoretical Biology*), was first rejected by a dozen other scientific journals.[6] Its huge significance comes only in part from the unexpected instance of hybridisation; the real clincher that makes this paper a classic is that the hybridisation in question produced a species that would go on to establish a whole domain of the tree of life – the eukaryotes.

The eukaryote branch of the tree of life contains all of the most complex organisms we see around us: the large, many-celled animals, plants and fungi; but also the great majority of the tiny, single-celled life to be seen, with the benefit of a microscope, whizzing around in a droplet of pond water. Eukaryote cells can be identified, most immediately, by the nucleus (*karyon*) that gives them their name. But their sophistication compared to the bacterial and archaeal prokaryotes goes much deeper. They are almost always considerably larger – a eukaryote is to a prokaryote what a blue whale is to a badger – and, alongside the nucleus, they have numerous 'organelles' (little organs) within their cells.

* At the time of her breakthrough paper still named Lynn Sagan from her recently ended marriage to Carl Sagan.

The most obvious eukaryote organelles are the small, submarine-shaped mitochondria, which act as miniature power stations 'burning' carbohydrates with oxygen to produce the energy-rich molecule adenosine triphosphate (ATP), which is used as currency in all the biochemical operations of the cell. Running a large and complex eukaryotic cell with its many genes depends utterly on the mitochondria for a constant turnover of new ATP; all the cells of an adult human together produce (and immediately consume) an extraordinary total of 70 kilograms of this molecule every day.[7] Plant and algal cells have a second, lentil-shaped family of organelles called 'plastids', which can take various forms with various functions, but the most familiar plastids are the chloroplasts that make green plants (and algae and therefore mint sauce worms) green. In a complement of the process that takes place in mitochondria, chloroplasts use the energy captured from sunlight to turn water and carbon dioxide into carbohydrates; oxygen is the delicious waste product of this reaction.

Eukaryotic cells – but not prokaryotes – also have a complex system of connected tubes and bubbles and compartments whose walls are made of fatty membranes. These act as production lines for various cell products, or as storage vessels, or as garbage disposal units. And many eukaryotes have their own (very distinct) equivalent of the bacteria flagellum, which sticks out into the liquid that surrounds the cell, wafting to and fro to cause movement.

Margulis's theory that this complex eukaryotic cell was born of the fusion of two distantly related species was in great part an attempt to account for the origins of at least some of this complexity. But it has its roots in the work of the nineteenth-century Russian biologists Andrei Famintsyn, Boris Kozo-Polyanski and especially Konstantin Mereschkowski.[8]

Mereschkowski was a deeply nasty man; it is difficult to convey quite how awful but perhaps better to get that out of the way before thinking about his contributions to the tree of life. He was born in 1855 in Warsaw (then part of Russia). His father, a reasonably important civil servant, seems to have been fairly unpleasant, disapproving of his son's decision to become a natural scientist, preferring him to become a lawyer. Mereschkowski studied in St Petersburg and then, in his twenties, had the opportunity to work at various Russian and European marine biology stations, earning his diploma (with distinction) and *Privatdocent* (licence to teach) along the way. He married in 1883 and then, in 1886, moved 'suddenly' to the Crimea.[9] In 1898, leaving his wife and toddler to fend for themselves, he again moved 'suddenly', to the USA, where he lived and worked under a false name. The sudden moves (and the alias) seem to have been an effort to escape the scandals of his paedophilia; scandals that finally caught up with him in 1914 when his past abuses became widely known – discussed in newspapers and the Russian parliament and culminating in court cases. At this point he fled again, this time to France where he lived for most of the First World War, before a final move to Switzerland. He was an enthusiastic and active racist, a eugenicist, as well as a paedophile, and, to complete the set, he had a messiah complex, writing a perverse book fantasising about his perfect world[10] and, on his death, leaving instructions for his imaginary disciples. Finally, in 1921, having run out of money, he killed himself in the ironically named Hôtel des Familles in Geneva using a somehow unsurprisingly ghoulish method. His body was found tied up and strapped to the bed; he had donned an elaborate mask contraption, and this, when his free hand opened a valve, had filled his lungs with an asphyxiating and poisonous gas.

The odious Mereschkowski's insight into the origins of eukaryotes was not, even at the end of the nineteenth century, entirely original, but he was undoubtedly its most forceful and thorough proponent. The theory concerns the origins not of the eukaryotic cell itself (although it clearly prepares the way) but of the tiny, lentil-shaped photosynthesising chloroplasts found in the (eukaryotic) cells of plants and algae. Mereschkowski proposed that chloroplasts are not structures that are manufactured by the plant cell itself – like the cell wall or the nucleus – but independent organisms — a species of photosynthesising bacteria (and therefore a prokaryote) that has made its home inside the eukaryotic plant cells. What he proposed, then, is that a plant is not one organism but two, that there is both a host plant cell and a guest chloroplast living in a symbiotic relationship. A more intimate relationship between two evolutionary strangers is hard to imagine.

This relationship of host and guest has been named 'endosymbiosis' – 'living together inside' – and it tells us that plant chloroplasts did not appear and evolve in the classic Darwinian fashion. Chloroplasts were not gradually evolved by plant cells; they arrived, *Deus ex machina*, in an instant. A proto-plant cell, incapable of photosynthesis, swallowed (but didn't digest) a fully functioning photosynthesising bacterium (a cyanobacterium), gaining the superpower of creating food from light. It is like the origin story of Spiderman – magical new powers gained at a stroke through the conjunction of Peter Parker and a radioactive spider.

Mereschkowski's evidence for this idea came in three parts. First were the clear examples, by then well known, of previously independent organisms fusing, one inside the other (we have already met the mint sauce worm with its algal guests). In 1867, the Swiss botanist Simon Schwendener had proposed that this kind of relationship was also true for the lichens – the familiar green-grey, yellow or orange crusts that slowly grow to cover gravestones, brickwork and trees.[11] Schwendener discovered that lichens are composed of two separate organisms, two separate branches of the tree of life living communally. The bulk of a lichen is a network of fungal cells; the second participant, living nestled amongst the fungal tissues, is a collection of algal cells, and these, because they contain chloroplasts, can photosynthesise and provide the fungal host with carbohydrate food (just as they do for mint sauce worms). Mereschkowski proposed that all plants are able to photosynthesise thanks to an equivalent symbiotic relationship with the photosynthesising chloroplasts. But in plants he saw a much more intimate association between the two cells: the photosynthesising bacterium living as a permanent guest not alongside but right within the eukaryotic host cell.

Mereschkowski's second strand of evidence for endosymbiosis came from the way that chloroplasts get passed from one generation of plant cell to the next. New chloroplasts are never formed from scratch (by the build-up of components in the cytoplasm of the plant cell) but are always and only formed by duplicating an existing chloroplast. This is as unexpected as a part of a human body – the arm for example – always being formed by the duplication of an already existing arm. This way of making new chloroplasts is surprising, yet it looks for all the world like the way in which new bacterial cells are made – a parent cell splitting in half to produce two daughters. This self-duplication is in fact a fundamental characteristic of being a cell – *omnis cellula e cellula*, 'all cells come from cells'.[12]

The third, and perhaps most direct piece of evidence came from the detailed similarities Mereschkowski discovered between plant chloroplasts and the cyanobacteria.* This was not an easy argument to make at the turn of the twentieth century, before the invention of the electron microscope, but the similar way in which they divide, their lack of a nucleus and, most obviously, their green colour and ability to photosynthesise were all used to make the link.

Mereschkowski's theory was one of those that floated around for the first sixty-odd years of the twentieth century, occasionally revived but mostly ignored, while people were distracted by genetics, biochemistry, the discovery of DNA and so on. The only important development during these otherwise fallow years was the expansion of the endosymbiosis theory (most persuasively by the American Ivan Wallin in 1925[13]) to include the origin of the energy-producing mitochondria that are found in all eukaryotic cells – the organelles that most clearly distinguish all eukaryotes from prokaryotes.

Lynn Margulis (née Alexander) was a remarkable person and scientist. Her obituary in *Nature* (she died in 2011) gives a hint of her character, beginning with her own claim that, from fourth grade, she had learned how to 'tell bullshit from . . . real authentic

* Cyanobacteria had long been misleadingly known as blue-green algae. They are bacteria and *not* algae (which are eukaryotes).

experience'.[14] She went to the University of Chicago aged just fourteen, graduating with her first (Liberal Arts) degree at seventeen. While there she met her equally charismatic first husband, the astronomer and famous populariser of science, Carl Sagan. After getting her MA in zoology and genetics in Wisconsin and her doctorate from the University of California, Berkeley, she moved in 1965 to Boston University with her two sons, her marriage to Carl Sagan having ended. Her career- and reputation-establishing paper was published soon after, in 1967.

Mereschkowski, as we have seen, was not entirely original, and neither, it follows, was Margulis. Her long and complicated paper, though, was so careful, so well evidenced and gave so complete an explanation of known facts that it became the first version of the endosymbiotic origin of eukaryote organelles to be generally accepted. Her theory tells us that there once was a cell, an unidentified proto-eukaryote, possibly already possessing some of the attributes of eukaryotes – a nucleus for instance – that desperately needed a way to deal with the poisonous oxygen that had begun to accumulate following the evolution of photosynthesising cyanobacteria. The antidote to the poisonous gas was found in a group of bacteria that had evolved the means to use oxygen to burn carbohydrates to produce energy. The proto-eukaryote latched on to this oxygen-using bacterium, becoming more and more intimately associated with it until eventually engulfing it entirely. The bacterium now lived permanently cosseted within the proto-eukaryote, consuming the deadly oxygen, all the while providing lots of lovely energy for the host cell. In Margulis's own words (in regrettably stilted academic-ese):

> the first step in the origin of eukaryotes from prokaryotes was related to survival in the new oxygen-containing atmosphere: an aerobic prokaryotic microbe [i.e. the protomitochondrion] was ingested into the cytoplasm of a heterotrophic anaerobe [i.e. one that gets its energy by eating stuff not by photosynthesis]. This endosymbiosis became obligate and resulted in the evolution of the first aerobic . . . amoeboid organisms.[15]

To represent the event that Margulis proposed on the tree of life, we are forced to picture the fusion of two extremely distantly related branches to make an entirely new (and ultimately rather amazing) branch – the eukaryotes.

The next step in Margulis's theory fitted with Mereschkowski's idea for the origin of chloroplasts – a second instance of endosymbiosis in which a species of eukaryote (a product of the first endosymbiosis) engulfed a photosynthesising cyanobacterium. The new branch would go on to give colour to our world – the greens and browns and reds of trees and grasses and seaweeds and the endless colours of flowers.

Margulis's strongest new evidence in support of her theory seems to have been the recent discovery that both mitochondria and chloroplasts had their own DNA – separate from the DNA in the cell's nucleus. This separate store of genetic information in organelles could be very elegantly explained if they had once been free-living organisms, independent of any eukaryotic host.

Resistance to endosymbiosis proved tough to smother entirely, its last gasp appearing as late as 1982 in a paper titled, 'Has the endosymbiont hypothesis been proven?'[16] The answer to this question was a tentative 'yes'. The final indisputable proof came from the discovery of copies of the universal small subunit ribosomal RNA gene hidden within the tiny genomes of both mitochondria and chloroplasts.[17] This gene, as is obvious from Woese's experiments comparing it in Eubacteria, Archaea and Eukaryota, has been inherited from LUCA by all of modern life. The existence of this gene in two forms in the two kinds of organelle proved that they are (or at least once were) living entities in their own right.

More exciting than this, by comparing the chloroplast and mitochondrial SSU rRNAs with the same gene from species right across Woese's tree of life, it ought to be possible to discover the identities of the two original bacteria that had been taken up by the proto-eukaryote cell to form the mitochondrion, and by the proto-plant cell to form the chloroplast. The sequences of the SSU rRNAs from chloroplasts have recently been shown to be most similar to those of a fresh-water cyanobacterium called *Gloeomargarita lithophora*; the freshwater habitat perhaps telling us something important about where this endosymbiosis happened.[18] Discoveries concerning the origins of the eukaryote nucleus and mitochondria will (in a later chapter) reveal to us what living species *they* most closely resemble and, hence, whence they came.

If we pause to think about it, the finding that our cells contain hitch-hikers is faintly unsettling. Mereschkowski, Margulis and other prescient supporters of endosymbiosis were able to think the (almost) unthinkable, that two of the biggest events in the evolution of complex life – the origins of all complex eukaryotes (animals, plants, fungi, ciliates, amoebae, foraminifera, coccolithophores and a great many similarly obscure single-celled organisms) and all photosynthesising eukaryotes (from seaweeds to mosses to Venus flytraps) – depended on the merging of entirely distinct creatures separated from each other by billions of years of evolution. We now know that the two lineages that combined to make the first eukaryote are twigs on the two very most distantly related branches on the entire tree of life. The host came from Woese's archaeal branch and the hitch-hiker from his bacterial branch. These are the two children of LUCA, and from the moment of their separation, these two lineages had coexisted, proliferating, evolving and diversifying, but separately, never meeting, never mixing – for more than 2 billion years. Their endosymbiosis makes the imaginary mating of a camel and leopard to make a giraffe look almost mundane.

What made this implausible event possible was the way in which it happened. Not in reality the instantaneous merging of two disparate genomes – like the mechanically improbable mating of camel and leopard – but a more gradual process of two organisms living beside each other gaining mutual benefits. The association slowly becoming more intimate and the species more codependent. One species, the mitochondrial ancestor, at some point being engulfed by the other but not digested, each of the two – host and guest – then gradually altering their genomes to adapt to and to benefit from the new situation. A process that must have taken millions of years to complete: a heartbeat in the grand scheme of evolution, but not quite instantaneous. Margulis's gentle, cooperative view of evolution has always been seen as lying in stark contrast to the mainstream Darwinian view of competition between individuals and between species for precious resources in a constant battle for the survival of the fittest. Both, of course, can be true.

Since their merger more than 2 billion years ago, mitochondria (and only slightly more recently, chloroplasts) have become intimately integrated into the biology and biochemistry of their host cells. While it was the existence of DNA in the mitochondria and chloroplasts, separate from the nuclear DNA, that confirmed their origins as independent organisms, the truth is that, like a bald man with a comb-over, mitochondria and chloroplasts are barely hanging on to their remaining genetic assets. Most bacteria have perhaps 4,000 or 5,000 protein-coding genes in their genomes, but a human mitochondrial genome now contains a mere 13. Mitochondria and chloroplasts have lost so many of their genes for two reasons. First, many of their genes simply became redundant – mitochondria and chloroplasts have no need to make most biochemical products, for example, because these are supplied by their much bigger host. But perhaps more interesting, from a tree of life point of view, are the genes that are still useful to the endosymbionts but have been transferred to the nucleus of the host to be incorporated right into the host's chromosomes.* These endosymbiont genes are now controlled by the nucleus, and this centralisation of control makes the symbiosis work more efficiently. This is the ultimate in codependency – a joint bank account –

* This transfer of genes was anticipated shortly after Margulis's paper of March 1967. In the June 1967 edition of *Nature*, the Norwegian Jostein Goksøyr builds on her points to make some prescient predictions for the fate of mitochondrial genes: '. . . the aerobic partner [the mitochondrion] must necessarily have lost a great part of its autonomy. On a molecular basis, the loss of autonomy must mean a loss of DNA. This DNA may have become incorporated into the nuclear DNA, giving the eucaryotic cell a still better control over its aerobic partner.' (Goksøyr, J. (1967). 'Evolution of Eucaryotic cells'. *Nature* 214:1161.)

and it means that the genomes of all eukaryotes are a chimera. Eukaryote genomes are a hotchpotch of genes whose origins lie in *both* founder members of the lineage. We are neither host nor guest but both at the same time. And if I were to pick one of your genes at random and use its DNA sequence to ask whether you are more closely related to the archaeal branch of life or to the bacterial, the result will depend entirely on whether I happened to select a gene that had its origins in the host or in the hitch-hiker.

The origins of mitochondria and chloroplasts from two lineages of bacteria are now clear (even if, with some inevitability, the precise identity of the closest relatives is still disputed). But what about the organism – the proto-eukaryote – that did the engulfing? If we could identify this beast, we could ask whether it already had some or all of the unique features of eukaryotes, which are lacking in living prokaryotes. Did the engulfer already have a nucleus with chromosomes, mitosis and meiosis, a complex system of internal pipes and vesicles, flagella and cilia? Was it already large? What did it eat? What was special about how it lived that made it ready and willing to enter into a partnership with an aerobic bacterium? What problem did it have to which the answer was endosymbiosis? The surprising identity of the engulfer amongst the diverse members of the archaeal domain has only just begun to emerge in the past few years, and it is something we will return to.

For the rest of her career, Margulis insisted on the outsize and (according to her) underappreciated importance of the merging of branches to the tree of life. Most evolutionary biologists now believe she went too far. Margulis started out a maverick who was vindicated but, in at least some of her beliefs, ended up at the crank end of the science spectrum. In 2009, she helped a British scientist called Donald Williamson publish a paper that, thanks to Margulis's membership of the august American Academy of Science, managed to bypass the careful scrutiny a paper is normally subjected to before publication. It proposed that the caterpillar stage of the butterfly life cycle evolved when an onychophoran or velvet worm (a distant relative of insects that we met briefly before) mated with the ancestor of the butterflies. Velvet worms have a squishy body, stumpy legs and an undeniable resemblance to a caterpillar. Williamson's idea was easily disproved because the number of butterfly genes that resemble those of velvet worms is precisely zero. The title of Williamson's paper was 'Caterpillars evolved from onychophorans by hybridogenesis'.[19] The paper written in response had the equally unambiguous title 'Caterpillars did not evolve from onychophorans by hybridogenesis'.[20]

Williamson's idea, then, was disproven by showing that butterfly genes bear no resemblance to velvet worm genes. But the same kind of test could also be used to uncover crossovers that actually did occur. Any foreign gene hiding in a living genome could be unmasked if a closely related gene could be found in a species from the foreigner's distant branch of origin. As soon as we amassed enough gene sequences from enough species, this experiment became possible, and it revealed something rather disconcerting for tree-builders. The jumping of genes (singly or even in groups) from one branch of the tree to another has been, and presumably still is, a fairly common occurrence. This raises the scary question of whether there even is a single tree of life to discover. Could it be that the only tree of life that exists is a tree of individual genes? Perhaps each gene can be followed on its species-hopping journey for a short while, but ultimately all the branches interweave into some sort of gigantic, impenetrable fishing net.

In this pessimistic view, species themselves might be viewed as being rather like Plutarch's Ship of Theseus, which 'was preserved by the Athenians down even to the time of Demetrius Phalereus, for they took away the old planks as they decayed, putting in new and stronger timber in their places'.[21] Is there a point at which so many planks have been replaced that this is no longer the Ship of Theseus? If most of the genes in a given species come from various elsewheres on the tree, does it make sense to speak of the lineage of that species?

The movement of a gene from one branch to another is called 'horizontal gene transfer' (HGT) – as opposed to the normal 'vertical transmission' of genes from parent to offspring. For reconstructing the tree of life, we need to know whether HGT is an interesting but rare quirk of evolution or something that is frequent enough to mess everything up. It turns out, thank goodness, that, while most genes do occasionally travel by HGT, by far the most likely fate of any individual gene in any branch of the tree of life is to be inherited vertically, faithfully remaining within the confines of the species' branch.[22] Perhaps equally helpful is the discovery that some genes are immune to HGT; there is a core set of genes – a small set admittedly – that are so essential to life and so tightly integrated in what they do with a host of other genes that they are effectively unchanging and irreplaceable. I like to think of the journey along a branch of the tree of life as being represented by a voyage on the Ship of Theseus from Athens to Crete. Even if some or even many planks get swapped out for others along the way, Theseus would nevertheless experience such a journey as an unbroken voyage on a single ship, and there will always be a few planks, somewhere on the ship – HGT-resistant genes – that will make it all the way to Crete.

An endosymbiosis as profound and permanent as that between the proto-eukaryotic host and its bacterial hitch-hikers is much, much rarer than the less intimate way of living together that we saw in lichens and mint sauce worms. But these are still beautiful examples of the melange of DNA coming from different places on the tree of life and coexisting in a single body. Taking the mint sauce worm as our example, we can enumerate the disparate contributions to its biology by studying its ribosomal RNAs. These core genes are ciphers representing each of the different genomes that the worm contains. Imagine taking one poor worm and grinding it up to find what different copies of ribosomal RNA we can find. We can see, first of all, that because it is a eukaryote, there must be one ribosomal RNA that is typical of the eukaryotic nucleus and a second ribosomal RNA that comes from the mitochondrial endosymbiont. But its green colour suggests another set of genes and genomes. The colour comes from a second eukaryote – an algal species called *Tetraselmis convolutae* which lives amongst the cells of the otherwise colourless worm. *Tetraselmis* brings with it three more ribosomal RNAs: the first from its own eukaryote nucleus; the second from its mitochondria; and the third from its chloroplast. In our ground-up mint sauce worm we would therefore discover a total of five different ribosomal genes, each revealing a contribution from a different branch of the tree of life. With these clues in hand, we can travel back in time to recreate the series of mergers that have forged these amazing creatures.

The mint sauce worm, *Symsagittifera roscoffensis*, is possibly the best known (or, at least, the most easily found) of the roughly 400 species that make up the otherwise lamentably little-known animal phylum that has been given (by me) the unfortunate mouthful of a name 'Xenacoelomorpha'. Where this phylum of worms belongs within the animal tree of life is the *casus belli* of a long-running dispute (war of the worms?) to which I will return. Like most arguments over the shape of the tree of life, it turns on the possibility that errors have been made (by me or by my frenemies) when building this part of the animal tree of life. It is the various causes of these mistakes – which are the bread and butter of my research – that we are going to explore next.

12

When Trees Go Wrong

IF ERNST HAECKEL'S oak tree is considered the starting point for the tree of life, then the project to find the correct place for every known species has already lasted nearly 160 years. Huge progress has been made, but the arguments nevertheless go on. This slow progress seems especially surprising when we remember that, even in the 1860s, Haeckel got much of the tree correct. It's even more puzzling when we remember the exploding availability of molecular data – billions of nucleotides from hundreds of thousands of species – and computers capable of crunching them. Why are there still jobs for phylogeneticists?

The general explanation is that, while big bits of the tree are easy to get right – it's easy to see mammals and birds are separate branches, and even easier to separate animals from plants and fungi – some parts, the bits that nerds like me are still arguing over, are much more confusing. Fortunately for non-nerds, these arguments are almost always caused by interesting bits of biology and give unique insights into the hows and whys of evolution.

Although the connection between its yin and its yang may not be immediately obvious, convergent evolution is a real mixed blessing for evolutionary biologists. It is the unifying phenomenon that lies behind all the errors we make when discovering the structure of the tree of life (remember swifts and swallows); but at the same time, it provides the treasured data that allows us to put our 'Just So' stories to the test. Why do leopards have spots and elephants trunks? Why do humans (uniquely) have a chin? And what does the size of our testicles tell us about monogamy? I will return to testicles, chins and the thorny details of convergent evolution, but first, let's look at the specific challenges we face when dealing with the shortest branches on the tree of life.

The short branches that lead to confusion working out the shape of the tree of life are not those right at the top leading to the leaves of the tree but those deeper inside the tree that separate two branches from one another. Their short length represents a lack of change and hence a corresponding lack of new group-defining characters that could help us distinguish the branches. To understand the problem, imagine we have just three species – a crab, a horse and a zebra – and we want to know which pair is the most closely related. This is clearly a very easy problem to solve, but the reason it is easy is what we need to grasp. This problem is easy because of the huge distance between the common ancestor of all three animals and the much more recent common ancestor of the horse and the zebra. A long branch between these two ancestors represents the accumulation of a correspondingly large number of characters (backbone, lungs, legs, hair, mammary glands, hooves, mane, etc.) inherited by the horse and the zebra but not by the crab.

The problem caused by short branches becomes clear when we consider the opposite situation, where the internal branch on the tree is very short. Let's take another set of species: a horse, a zebra and a donkey. Which pair is the most closely related? Not so easy is it! The correct answer (probably – it really isn't an easy problem) is that zebras and donkeys are each other's closest relatives, but so little time passed between the existence of the horse/zebra/donkey ancestor and the zebra/donkey ancestor that the zebras/donkeys have had almost no time to accumulate characteristics that clearly distinguish them from the horse. In fact, experts are only just able to distinguish donkeys and zebras from horses on the basis of tiny differences in the shapes of the whorls of enamel on their teeth.[1]

If we think of short branches in terms of changing gene sequences instead of morphology, we find the exact same problem.

The likelihood of a mutation happening correlates with how much time passes (more time means more mutations accumulating). If we took any gene from each of our three equid species and looked for differences between them, we should expect very few to have accumulated in the branch separating the common ancestor of all three animals and the ever so slightly more recent zebra/donkey ancestor.

Short branches occur all over the tree of life, and the problem can get much more difficult when more than three branches – sometimes many more – all split apart from each other in a very short period of time. For example, I left out several living and tens of extinct species of equid (onagers, kiangs, kulans, Mongolian, Syrian, Indian and Somali asses, several species of zebra, the recently extinct quagga and so on) from the study described above. These closely related species branch out from each other in close succession in a pattern called a radiation, the most famous of which may be the one that produced fifteen closely related species of Darwin's finches on the Galapagos Islands.[2] The radiation of the Galapagos finches occurred when individuals from one or a few species arrived on the islands, probably blown the hundreds of miles from the South American mainland by a storm. Once there, the tiny founder population rapidly diversified into different-looking forms living in different habitats on different islands and eating different foods. The niches they now occupy are most obvious in their very different beaks:* large and chunky in *Geospiza magnirostris* (*magni rostris* meaning 'big beak'), which is able to crack nuts; sharp in *Geospiza septentrionalis* (the 'vampire ground finch'), which pecks the skin of other birds to draw blood to drink; and long and pointy in *Geospiza difficilis*, which feeds on insects and snails.

* 'The beak of Cactornis is somewhat like that of a starling; and that of . . . Camarhynchus, is slightly parrot-shaped. Seeing this gradation and diversity of structure in one small, intimately related group of birds, one might really fancy that from an original paucity of birds in this archipelago, one species had been taken and modified for different ends.' (Darwin, C. (1860). *The Voyage of HMS Beagle. A Naturalist's Voyage Round the World.* John Murray, London.)

Radiations such as the one that produced the finches are a phenomenon dearly beloved of evolutionary biologists. They seem to occur when new opportunities open up for a founder species – the Wild West of an unoccupied island for the finches. Probably the most majestic of all these new opportunities was the opening of the East African Great Rift Valley, which produced the string of Great Lakes – Victoria, Tanganyika, Malawi, Turkana and so on. These new lakes were the opportunity for the explosive radiation of cichlid fishes, perhaps 2,000 species of every size and shape and habit, each adapted to its own niche as hunter, scavenger, herbivore or molluscivore: some are brilliantly colourful, some drab and some cunningly camouflaged; some are found in the shallows, some swimming in open water and some hugging the bottom in the depths of the different lakes.[3] The relationships between the species produced in these explosions of creation are all but impossible to know for certain.

A possible solution to the short branch problem produced by radiations is to collect information from so many characters (almost all of which will *not* have had time to change in the brief moments of the radiation) that we can be certain of picking up at least a handful of characters that *have* changed and can tell us about species relationships. This gathering of huge numbers of characters is much easier now that we can use DNA or protein data. In the case of the horses and their relatives, it required an analysis of a huge number of genes (more than 20,000 genes and 31 million DNA letters[4]) to find enough rare changes to tease their relationships apart. But for the Galapagos finches, even entire genomes' worth of data give inconsistent and confusing answers regarding the relationships between species.[5]

Although the problem of short branches can generally be dealt with using big datasets, the scarcity of helpful, honest characters makes it much easier for other sources of error to exert their terrible influence. This is where we will venture next.

Dishonest or misleading characters are produced by a common evolutionary phenomenon called homoplasy – we have met this before in its most common guise of convergent evolution. 'Homoplasy' refers to two similar or identical characters that are observed in two branches of a tree of life, but which (in contrast to homology) were *not* inherited from their common ancestor. A homoplastic character, in other words, is a character that has evolved completely independently in two branches and fools you into believing the branches to be more closely related than they are. A blatant example (for which there is no real danger of imagining homology or close relationship) would be the wings of insects and of birds – evolved completely independently by both groups to allow flight. We have seen far more subtle examples too, such as the similar forms of swallows and swifts.

The word 'homoplasy' was coined by E. Ray Lankester, the illustrious first occupant of the chair I now hold at UCL. A disciple of Thomas Henry Huxley ('Darwin's bulldog'), Lankester himself was a pugnacious defender of evolution and the scientific method in general, earning himself the nickname of 'Huxley's bulldog'. His stout defence of science had notable and sometimes surprising successes, including the unmasking and prosecuting of an American psychic. 'The spirit medium is a curious and unsavoury specimen of natural history, and if you wish to study him, you must take him unawares, as you would any other vermin.'[6]

The generally wonderful Lankester had feet of clay, being one of the eminent Edwardians who fell for the famous Piltdown Man hoax – a supposed fossil of the 'earliest Englishman', found complete with a bone cricket bat, whose skull was that of a modern human and whose jawbone came from an orangutan.[7]

Lankester's simile for homoplasy, of two distant groups of cavemen who independently invented stone axes or dugout canoes, still works well. These two inventions, axe and canoe, though not inherited by the two groups from a common ancestral population that had already invented them, would nevertheless be substantially similar because they fulfil the same roles and are made from the same materials. As Lankester's example illustrates, homoplasy, especially at the level of phenotype, arises when two independent branches of the tree come up with a similar solution to a common problem (birds' and bees' wings). Analogous rather than homologous.

The Tasmanian wolf (aka the thylacine or Tasmanian tiger) is a classic example of convergent evolution and also one of the saddest stories of a recent extinction. The youngest Tasmanian wolf skeletons found on mainland Australia date from 3,200 years ago,[8] but the species survived in the wilds of Tasmania right into the twentieth century. European settlers were the inevitable cause of their final demise; they were persecuted as a danger to sheep and chickens (with bounties paid to hunters) in just the same way that wolves were once driven out of Europe. A film from 1933 shows a male Tasmanian wolf pacing its cage in Hobart Zoo, and it is a terribly sad thing to see. The very last of the species, a female, died in captivity in 1936.

Even if the 'wolf' moniker isn't quite right – they are smaller and striped and have shorter fur – Tasmanian wolves really are physically very similar to wolves, foxes, dogs and so on. Their bodies are shaped like a pointer's – slim with longish legs that, just like a dog's, end in a foot raised off the ground, and with clawed paws at their ends. They have the forward-facing eyes of a predator; alert, slightly rounded ears like a corgi's; and a long muzzle that ends in a black nose. Their skulls in particular are almost indistinguishable from those of dogs and wolves, being beautifully designed for eating meat – big jaws, big teeth and big bones for big muscles. Closer inspection has shown that the Tasmanian wolf skull finds its nearest match not in the skulls of wolves but in those of jackals, coyotes and foxes, ironically suggesting (because form follows function) that, far from preying on sheep, the thylacine favoured smaller prey animals such as mice, rats and rabbits.

Despite the extraordinary similarities, we know for sure that Tasmanian wolves are far from being close relatives of the dog family, but it is only when we compare *all* their many characteristics that we discover this. The Tasmanian wolf is in fact a marsupial, as is most evident from its possession of a pouch for carrying young. It is (was) a relative of kangaroos, quolls and koalas. Dogs, jackals, foxes and wolves, in contrast, are placental mammals (babies grow to full term inside their mother), and this tells us that dogs and co. are much closer to humans, bats, cows and whales than to the Tasmanian wolf. The broad problem this kind of homoplasy brings, then, is to fool us into believing two species (or any two branches on an evolutionary tree) are more closely related than they really are.

There is an equally problematic second form of homoplasy that can be called 'reversion' or 'character loss', which happens when a species loses some, or even many, of the characteristics that define the branch of the tree that it belongs to. An example of the effects of this sort of character loss can even be seen on Ernst Haeckel's first tree, on which we find a branch labelled 'Himatega', which Haeckel places close to the molluscs.[9] 'Himatega' is Haeckel's name for the sea squirts – rather unappealing creatures which resemble bags of jelly with two pipes sticking out, often encased in a flexible but tough coat called a tunic (they are also known as tunicates). They live what seems a very dull life attached to rocks or piers, making a humble living sucking water in through one pipe, filtering out any food particles with an internal sieve that fills most of their body, and squirting the waste water out of the second pipe – hence the name.

The similarity of this bag of jelly to some of the less glamorous molluscs (from Greek *malakós*, meaning 'soft') is obvious, but it turns out that the simple mollusc-like body of sea squirts misled Haeckel (and Linnaeus and Aristotle before him). Their real identity was revealed in 1867 by the Russian biologist Alexander Kowalevsky, who was the first to study the whole life cycle of sea squirts.[10] The sea squirt hatches from its egg in a form called a tadpole larva, which looks completely different to its mollusc-like adult body. Kowalevsky found that sea squirt tadpole larvae share important similarities (homologs!) with our own branch of vertebrates – most significantly of all an equivalent of our backbone called a 'notochord'. Haeckel quickly recognised Kowalewsky's discovery in his popular book *The History of Creation* (1868): 'The early stages of the [tunicates] possess the beginnings of the *spinal marrow* and the *spinal column* . . . the two most essential and most characteristic organs of the vertebrate animal.'[11]

The reason that Haeckel and others were initially mistaken about the place of sea squirts on the tree of life is that adult sea squirts have discarded almost all their obvious vertebrate characters, including the backbone/notochord. They presumably disappeared down this regressive evolutionary hole when they adopted their boring but successful new way of living, fixed to a rock filtering seawater – they no longer had any need for a backbone.

The evolution of simplicity from complexity, and more generally the loss of characters, is not something we tend to focus on when we think about evolution, which is almost synonymous with the steady accumulation of new or more sophisticated characters. The loss of characters and of complexity is, however, a central part of the evolutionary process: just think of our own hairless, tailless, gill-less selves. Losses occur at all levels of biology from big bits of a body (notochord/tail) right down to individual genes. While the loss of characters is common, there are a few branches on the tree of life that, like the sea squirts, have taken this to extremes. These oddballs are amongst the hardest to place on the tree of life.

Even sitting in a perfectly clean and utterly empty room, you would be a very long way from being alone. You are, in fact, the home planet of a plethora of aliens. Your hair, skin and guts (especially your guts) are host to trillions of bacteria, some beneficial, many harmless and some downright nasty. Your intestines and blood and cells and organs can be infested with a rogues' gallery of single-celled eukaryotes (responsible for the horrors of malaria, giardia, amoebic dysentery, athlete's foot, thrush, ringworm, sleeping sickness and more). And a great deal of human misery is caused by larger organisms – parasitic animals that live on and within planet human: tapeworms, roundworms, flukes and schistosomes; lice and ticks; fleas and botflies; chiggers, tongue-worms and crabs. Even the healthiest and most fastidious of us is home to bands of tiny creatures such as the eyelash mite *Demodex* which lives (harmlessly) like a hermit in the cave-like follicles of our eyelashes and in the pores of our skin. This brief census tells us that parasites (not just those that afflict humans) hail from many different branches, each one an independent experiment in this odd way of life.

If we focus simply on the larger of these parasites and examine the biology of the *animals* that infect us, there are some collective weirdnesses that jump out. All of these animal parasites have had to adapt to a very specific way of life, and they have often taken parallel paths. One of the commonest stories of parasite evolution describes a series of losses and reductions: animals become tiny and simplified; their brains shrink; some lose their mouth or anus or have no intestine at all; their limbs, if their ancestors had them, can become stumpy or entirely lost; their eyes are often blind; their complement of genes is often greatly reduced; and so on. Trying to identify the relatives of these depleted creatures can resemble the vain search for traces of the hobbit Sméagol in the hideous form of Gollum.

Perhaps the most amazing champions of simplification and loss among the many animal parasites are two little-known groups called Dicyemida ('two germs' – referring to the existence of both sexual and asexual reproduction) and Orthonectida ('straight swimmers'), both of which make a roundworm appear like a sophisticated and complex machine.

The orthonectids were first described in the late 1870s by the French zoologist Alfred Giard (who has the questionable honour of having the nasty parasite *Giardia* named after him).[12] Giard found different but clearly related species of orthonectid parasitising two very different groups of animals: the brittle stars (relatives of starfish) and the nemertean worms (which include the longest animal of all, the aptly named *Lineus longissimus*, which, although they are hard to measure as they are quite stretchy, may grow up to 55 m long, almost twice as long as a blue whale). The orthonectids are tiny – a little over a tenth of a millimetre long (and so about 550,000 times shorter than *Lineus*) – and their entire body is made up of just a couple of hundred cells. In some species, the female nervous system comprises just four cells; the males, sad to report, are even stupider with two.

The dicyemids were described and named in the middle of the nineteenth century by the Swiss scientist Albert von Kölliker, who found them living in the kidneys of an octopus.[13] Since this time, a little over 100 species have been described, and every one of them likewise lives in the kidneys of a cephalopod – octopus, cuttlefish, squid or nautilus. Dicyemids have even fewer cells than the orthonectids, and descriptions of the dicyemids inevitably involve a list of characters that they do not have, most of which it might seem sensible to have hung on to: no organs or tissues; no internal body cavity; no gut; no nerve cells at all – even stupider than the orthonectids!

For most of the years since their discovery, orthonectids and dicyemids have been lumped together in a larger group, a phylum called Mesozoa. This name – which translates as something along the lines of 'intermediate animal' – gives a big hint that zoologists believed them to be a very early branch on the animal tree of life, fitting somewhere between Protozoa ('first animals', i.e. single-celled proto-animals) and Metazoa ('after the animals', i.e. proper animals like insects, molluscs and fish). In essence, the mesozoans were widely considered to be the living descendants of a very early stage of animal evolution.

This common view of these simple animals has proved to be misguided. Giard was surprisingly modern when he remarked:

> The Orthonectida are parasitic animals, and we must take into account the retrogression which this kind of life may have brought about in their structure. An organisation which we consider as one of primitive simplicity is very possibly simple only in consequence of degeneration.[14]

The same may, of course, be said of the even simpler, parasitic dicyemids.

Giard has since been proven right. It turns out that neither of these extraordinarily simple worms is an early branch off the main lineage of the animals. Both are related, in ways that have been rather unclear, to a branch that contains a whole zoo of much more complex species – including annelid worms, molluscs, flatworms and nemertean worms. These relatives of the orthonectids and dicyemids are large; they have a proper gut; they respond to their environments with complex behaviours controlled by complex nervous systems made of tens of thousands of neurons; they move and hunt and hide and mate using systems of muscles made of thousands of muscle cells. If dicyemids and orthonectids really are relatives of these big and complex animals then they must, in becoming parasites, have lost almost all of this complexity.

A few years ago, I worked with two colleagues – PhD student Dr Helen Robertson and postdoctoral fellow Dr Philipp Schiffer – to try to narrow down where on the animal tree the orthonectids and dicyemids belong. Our first idea was to look in their DNA for a complex and unusual character that we knew was only found in the group of complex animals that includes the annelids, molluscs and flatworms. This character takes the form of the unique make-up of a protein (one that is found in all eukaryotes) called NADH dehydrogenase subunit 5 (nad5 for short).

In most branches of the animal tree, the nad5 protein has changed very little, but in the branch that leads to annelids, molluscs, flatworms and company, it has accumulated a whole series of changes. Most of these changes are in the sequence of its amino acids, but the length of the protein has also changed, and there are even a few amino acids in the middle of the protein that have simply disappeared. This changed version of the nad5 protein is found only in this group of animals, and it stands as clear proof that they are all related to one another. As a character this altered nad5 is no different from the mammary glands that define mammals or the acorns that define oak trees. As for the dicyemids and orthonectids, we reasoned that if we could show that they had this unique version of nad5, it would be the killer evidence that these super-simple parasites really do belong to this group of complex animals.

This started out as a very basic question – if we could find the relevant gene in the DNA sequence of their genomes (which had been recently published), we would get our answer pretty much immediately. When we looked at the sequence of amino acids in their nad5 proteins it was obvious that both dicyemids and orthonectids had the exact same unusual version found in the annelid worms, molluscs and so on. The series of changes needed to make this transformation were so numerous and so improbable that there really was no plausible way that the two weird worms could have chanced upon them independently. They must have inherited the unusual version of nad5 from the same ancestor that had bequeathed it to the annelids, molluscs and so on. Our discovery showed that, rather than being simple because they are an early offshoot of the animal tree, dicyemids and orthonectids have *become* simple by the wholesale loss of characters through evolution.

We went on to study more of their genes, and we found that the minuscule orthonectids are one minor twig on the mighty annelid worm branch. Orthonectids, in other words, are relatives of the large and complex earthworms, lugworms, leeches and their relatives. And if annelids don't sound particularly complex to you, compare the brain of a leech, which has 10,000 nerve cells,[15] to the two-cell nervous system of an orthonectid. The simplification really is extraordinary; the reduction in neuron capacity, for example, is equivalent to a branch of humans evolving a brain the size of a grain of salt.

The dicyemids, orthonectids and sea squirts show how losing characters can cause errors in placing branches on the tree of life: when a member of a certain branch loses a character that otherwise defines that branch, there is a danger that we will be fooled into believing that it belongs elsewhere. This can be thought of as equivalent to convergent evolution (swallows and swifts, thylacines and wolves) – these simplified animals have reversed evolution and in so doing have converged on the characteristics of their ancestors.

13

Problems with Genes

Both forms of homoplasy – convergent evolution of new characters and loss of characters – are very common across the tree of life, typically scattered at random across the branches of the tree. These same problems also appear, albeit in a slightly different form, when we use DNA and amino acid sequences to build the tree.

There are two ways in which homoplasy manifests itself that make molecular data subtly different from morphological data – the first is the limited alphabet of molecular data (versus the endless avenues for morphological change). For each and every character (nucleotide) in a gene, there are only four states it can adopt. If the first nucleotide in a gene is an A (Adenine), there are only three different new states it can assume following a mutation: C, G or T (Cytosine, Guanine or Thymine). And if the same nucleotide were to mutate in another species, the same limited choice of outcomes is available. What this means is that if two distantly related species happen to change the same nucleotide in the same gene, there is (and I am simplifying in several ways here) a one in three chance that they end up changing to the exact same new nucleotide. It is as if, when birds were evolving a new shape of beak, they had only four kinds to choose from – finding the same beak in unrelated birds would be quite common.

Convergence is also likely, though slightly less so, in amino acids. Assuming again, for simplicity, that all changes are equally likely (they aren't), if the same amino acid changes in two unrelated species then there is a one in nineteen chance that they end up identical. Unlike my imaginary beaks, almost no morphological character is as constrained as molecular characters are in what changes they are allowed to experience; a careful examination of apparently identical morphological characters will almost always reveal differences, even if they are incredibly subtle. The small alphabets of biological molecules conspire to make *randomly occurring* convergent changes in the molecules of two species rather frequent.

The second way that evolving genotypes differ from evolving phenotypes pushes in the opposite direction to the problem of limited alphabets. The countless possible ways to change an organism's genome in order to alter its phenotype mean that two organisms will almost never evolve the same *phenotypic* character using the same exact changes to their genes. A flamingo and a giraffe, for example, have certainly not changed the same nucleotides, amino acids or even genes in order to evolve their long necks.

A simple illustration of this truth can be found in the unusual blue eggs found in some prized breeds of chicken (the Araucana chicken from Chile and the Dongxiang and Lushi chickens from China). The genetic basis of this unusual character has been studied in both the Chinese and the Chilean chickens, and it turns out that both breeds have evolved blue eggs by mutating the exact same gene. The gene has the rather prosaic name of SLCO1B3, and its normal function is to move coloured bile salts across cells. In both chicken breeds, the mutations that are responsible for the blue eggs cause the switching on of SLCO1B3 in the uterus of the chicken where the eggshell is being made, with the result that the blue-green bile salts end up in the eggshell.[1] When the DNA sequences of the SLCO1B3 genes were studied, however, it became clear that the mutations in the two breeds affected completely different parts of their SLCO1B3 genes. The entirely different paths they have taken to converge on an identical blue egg show beyond question that they achieved this independently. The message here is that where a common selective pressure may bring about convergent evolution in the *phenotype* of two organisms, at the genetic level, species almost always continue to diverge.

The overall outcome that emerges from the mix of these two different forces is that while convergent changes will certainly occur (due to the small alphabets of DNA and proteins), these changes should normally be randomly scattered across the branches of the tree. And while convergent changes will occur at the level of individual DNA letters and amino acids, whole sequences are very unlikely to converge to become misleadingly similar by chance; in fact, as long as we look at enough nucleotides or amino acids, they are actually bound to become less similar overall as time passes.

In the 1990s, a spat over the shape of the insect tree of life featured claims of convergent evolution of both phenotype and genotype. The argument has a surprising link to the four-winged mutant fruit fly we met in a previous chapter. To refresh your memory briefly before I get to the point: while most insects have two pairs of wings, all flies (those without mutations at least) have just a single pair; evolution has converted the fly's second pair of wings into little dumbbell-shaped organs called halteres thought to help the insect balance in flight. We also met in passing a second group of insects called Strepsiptera ('twisted wings'), and these also have a single pair of wings and a pair of halteres. Could their similar-looking structures be homologous, inherited by both groups from a common ancestor that also had halteres?

This idea was endorsed by the very first studies that used molecular sequences to understand the insect tree of life; these first trees grouped the flies and the twisted-wings together on a single branch, and this close relationship seems to tell us that the halteres that the two groups possess must indeed be homologous – inherited from their apparently close common ancestor. There is, however, a really intriguing problem with this sensible-seeming conclusion: while the halteres of flies have replaced the hind wings, the halteres of the twisted-wings occupy the place of the forewings. The explanation for the different places of halteres on the bodies of flies and twisted-wings (published in *Nature* and taken, I remember, very seriously at the time[2]) was that halteres had evolved a single time in the ancestor of the two groups (that halteres were therefore homologous) but that a series of mutations had swapped the wings and halteres around in the twisted-wings, as if our arms and legs swapped places. Very odd, but this was the heyday of the four-winged fly. Similar transformations (artificially induced ones), such as in the antennapedia mutant, which had swapped the antennae on a fruit fly's head for a pair of legs, made this an entirely plausible idea.

Ideas of such drastic and sudden changes in an animal's body are usually frowned upon as implausible 'hopeful monsters'.[3] This neat story relies, remember, on there being a close relationship between the two groups. The problem is that there is no such relationship. The tree of insects from the mid-1990s contained a glaring error: flies and twisted-wings are not close relatives at all.

Once the error was discovered, it was shown (just as the morphologists had said all along) that the twisted-wings are in fact most closely related to the beetles. And this does makes sense, as they both have front wings that have been converted into a new structure – the halteres in the twisted-wings and the elytra (the smooth carapace that covers up the delicate hind wings) in the beetles. Order restored. Hopeful monster slain. But what caused the fatal error in the tree? And have we learned how to avoid it?

The problem with the insect tree, called 'long branch attraction', was first discovered and explained by Professor Joe Felsenstein (now Professor Emeritus at the University of Washington in Seattle). Felsenstein is one of the gurus of molecular tree building. I met him (he certainly won't remember) at a course that I took when I was a PhD student in 1991. I remember him as absolutely terrifying, with a big black beard and scary intellect (though a quick Google search for recent pictures shows a rather friendly looking man with a broad smile). The course was held at the beautiful Woods Hole marine laboratory on the heel of Cape Cod in Massachusetts. The exact date of the course is easy to check; it was the week that Cape Cod was hit by Hurricane Bob. On 20 August, the *Washington Post* reported: 'Hurricane Rakes New England . . . "All roads on this island are impassable," said Police Chief Timothy Rich of Chilmark, a small town on the southern edge of Martha's Vineyard. "Trees are down everywhere. Power lines are lying in the road."'[4] As this extraordinary natural phenomenon raged outside, uprooting trees, piling yachts onto beaches and flooding the car park, our course continued inside, even power cuts proving insufficient to stop the lecturers from lecturing. The show must go on.

Long branch attraction is the most important of a family of related problems that afflict our trees when characters change in different ways on different branches. As was true of the fly and twisted-wing example, we encounter the long branch problem (we enter what is actually called the 'Felsenstein zone'[5]) when some branches on the tree evolve more quickly than other branches.

Felsenstein's famous paper in which he described long branch attraction was published way back in 1978, but its lasting importance is obvious in the roughly fifty citations it still receives every year (most scientific papers are cited fewer than five times *ever*).[6]

The paper describes an evolutionary tree that relates four species: Apple and Banana are related to each other on one side of the tree and Yak and Zebra are related on the other. The branch in the middle that separates A and B from Y and Z is (implausibly from my otherwise hopefully easy-to-follow example) short, which tells us there are very few changes that differentiate A and B on one side from Y and Z on the other.

The long branch attraction problem is going to arise because the four species have evolved at very different rates. Species A and Y have evolved very quickly, while species B and Z have barely changed at all. The large number of changes that have accumulated along the two fast-evolving (i.e. long) branches mean that both of them have evolved to become more and more different from their closest relatives (A has evolved away from B; Y from Z).

The error in working out their relationships occurs because, very, very occasionally, the exact same nucleotide change will happen on both of the long branches by chance convergence. These changed nucleotides are now characters the two fast-evolving species share (and which are not found in either of the slow-evolving species). But these occasional shared characters found in both A and Y were not inherited from a common ancestor.

The long branch attraction error results from the fatal combination of the very small number of changes that *correctly* separate A and B from Y and Z and the larger number of convergent changes *wrongly* telling us that the two long branches A and Y are linked. The result is an incorrect tree where the long branches are grouped together. Hence 'long branch attraction'.

Long branch attraction is a horribly complicated subject whose further subtleties I am going to deal with by ignoring them completely: the one thing we really need to know is that, as we saw with the long branched flies and twisted-wings, it causes fast-evolving species, even if they are not closely related to one another, to be grouped together on the tree as if they were.

For all its irritating ambiguity, convergent evolution is an important part of the story of evolution – and occasionally it is the grit that makes the pearl. The value of convergent evolution comes from its ability to tell us why certain characters have evolved. Darwin showed us that new characteristics evolve because they give some members of a species an advantage over the others. Often, but certainly not always, we can make an educated guess at what that advantage is – the genre finding its apotheosis in the *Just So Stories* of Rudyard Kipling. That we are fairly self-obsessed in this regard can be seen in the predominance of questions we ask about all that is special and unusual about humans, from the sacred to the profane. Why do human cultures believe in gods or an afterlife? Why do we walk on two legs? Why do men have nipples? Why do women have orgasms?

There's quite a lot of this sort of speculation about aspects of sex, it has to be noted, and so, taking up this baton with enthusiasm, we are going to ask just why the testicles of the Abyssinian black-and-white colobus monkey (*Colobus guereza*) weigh just 3 grams while those of the rather similar-looking bonnet macaque (*Macaca radiata*) weigh a whopping 48 grams.[7] And we are going to use the phenomenon of convergent evolution to answer this question.

The explanations we might think up for why a character such as big (or small) testicles has evolved can, of course, be very plausible, but usually we are just guessing (and this is not how we do science!). If we were asking such questions about a lab organism – a fruit fly, a nematode worm, a mouse or a yeast – we might be able to do some experiments to test our guesses. We could manipulate the character and observe the effect to see what it normally does – pulling the wings off a fly, for example, or mutating a gene – but these involved experiments are impossible for almost all species. We would very much like to have another, simpler way of testing our guesses, one that can be broadly applied across the tree of life.

There is one kind of scientifically kosher experiment we can use to answer the question of why something evolved. This method depends on finding multiple examples of the character's evolution and then asking whether it always goes hand in hand with an explanatory event or cause. This is like testing the theory that smoking causes lung cancer not by referring to a single uncle who was a smoker and got cancer but by looking at lots of smokers and lots of non-smokers and seeing whether there is a consistent correlation. For our purposes, this method requires us to find several separate examples of the evolution of the character we are interested in and, as you may have guessed, this means that we are now hoping to find cases of convergent evolution.

The Abyssinian black-and-white colobus is a striking animal; one of its alternative common names is the mantled guereza, which refers to the cloak-like fringe of long, silky white hairs that pop out in contrast to the short black fur that covers the rest of its body. Their heads are even more distinctive, the ensemble not unlike the headwear of a nun: there is a grey patch in the middle that covers the muzzle and eyes; this is surrounded by white cheeks, chin and eyebrows; and all of this is surrounded by black, including the top of the head where they have a pair of furry black lobes rather similar to buttocks. They live in small groups of ten to fifteen that consist of a single male, a harem of females and a handful of their young. The young males all leave their family group at adolescence to try, eventually, to establish a harem of their own. The dominant male colobus has fought for unhindered and easy access to his females; like a red deer stag, he maintains his position and his territory with a dawn chorus of roars.[8]

Bonnet macaques (*Macaca radiata*) have a very different way of loving. They are common in south-western India and look to me like the Platonic ideal of a monkey: short fur, big ears and a long tail; 'bonnet' refers to the tufts of longer hair on the top of their heads styled into a middle parting like an Edwardian stockbroker. Macaques live in troops of about thirty individuals with roughly equal proportions of males and females. They are both polygamous and polyandrous, meaning that males mate with multiple females and females with multiple males; males occasionally mount each other as well.[9]

The very different mating systems of the two otherwise quite similar animals – the monogamous harems of the colobus versus the free love of the macaques – is the difference that has been proposed to explain the sixteen-fold difference in the size of their testes. Promiscuous macaques, the theory goes, need to produce lots of sperm to fulfil the requirements of the multiple matings (and even more so because, in the race to the ovum, their sperm are in direct competition with those of other males). The colobus male has no such competition for mating with his harem and a small number of sperm is plenty to ensure the females' eggs are fertilised. The link between promiscuity and testicle size seems a very sensible explanation, but might there be others that are equally plausible? Maybe testicle size is a consequence of more protein in the macaque diet. Maybe macaques have bigger testicles to produce more testosterone rather than to produce more sperm. Like the plausible association between smoking and cancer, we could strengthen our belief in a link between large testicles and promiscuity (strictly speaking this is 'sperm competition') if we found multiple instances where the two things coincide.

One thing that we shouldn't do is rely on simply looking at lots of species of macaques – it turns out they are all promiscuous with large testicles – because each species has almost certainly inherited these two characteristics from their common ancestor. We can only count this as a single observation – the equivalent of a single smoker with lung cancer. If we are to answer this question convincingly, we need to look at lots of different groups of mammals. In other words, we are interested in finding lots of separate examples of convergent evolution of the characteristic we are interested in.

When this more nuanced approach is taken, our initial hunch is confirmed: right across the tree of mammals, males of polygamous species have proportionally larger testicles than males of monogamous species. The relationship between these two characters is even consistent enough to tempt us into making informed guesses about the sexual behaviour of any mammal based on its testicle size. Dolphins and porpoises, for example, have huge testicles, about 1% of their total body mass (the equivalent in a human would be testes weighing close to a kilo), suggesting a high degree of promiscuity. While obviously difficult to study in the wild, at least for spinner dolphins this prediction seems to pan out – spinners have been observed to engage in mass mating events called wuzzles (orgies by any other name).[10] Human male testicles – in case you are wondering – are somewhere in the middle; you can take from that ambiguity what you will! We can find an interesting contrast if we ask why humans have evolved a chin. Because we are the only animal that has a chin, there is no example of convergent evolution, and this means that the reason we evolved a chin (if there is one) will forever remain a mystery.[11]

Convergent evolution has turned out to be a double-edged

sword – both causing errors in building the tree of life and showing us why certain characters evolved. The next chapters will show that the tree of life can allow us to ask not only 'how?' and 'why?' but also 'when?'. We can dare to attempt this because much of evolutionary change occurs at a more or less steady rate, changes in the sequences of DNA and proteins in particular ticking away like a slightly unreliable grandfather clock. If we can work out how fast this clock is ticking, we can use it to set the coordinates for our time machine. An ability to add a timescale to our trees will let us place evolutionary events in the timeline of the ancient earth.

14
Rock of Ages (or Ages of Rocks)

THE TINY TOWN of Fortune lies on the north shore of the barren, lake-pocked Burin Peninsula, part of the Canadian island of Newfoundland. An inlet, protected by a spit of rock, makes the natural harbour that Fortune is built next to. On the rocky shoreline to the west of Fortune, a cautious scramble on the low cliffs will reveal a surprising addition to the rocks: what looks like a small, metal plaque on the cliff face. This is not in fact a plaque but the visible head of a long, bronze stake that has been hammered into the rock; it is a rare object known as a 'golden spike'. The golden spike's enormous significance is as a marker of a revolution that was happening on the earth at the exact moment that the layers of sedimentary rock in which it rests were being laid down.

The arguments over where the spike should be located began in 1969, and it took until the early 1990s for palaeontologists and geologists around the world to agree finally on just where it should go (there were several other candidates).[1] It is one of just a few dozen such spikes hammered into rocks around the world, each one marking the exact point in the geological record where the ancient earth transitioned from one time period to the next.

The land that Fortune was built on was not always bathed in cold subarctic seas; 550 million years ago, the layers of rock were being deposited as sediment on the edge of a land mass that sat squarely on the equator. The spike pinpoints the moment that something wonderful happened in these warm seas: a sudden transition in animal life from simple to complex, from sedentary blobs to active, curious hunters and seekers of food. The first appearance of bilaterian animals that we can recognise as fundamentally similar to ourselves.

First, we see the traces of *Treptichnus pedum*, the first bilaterian in the fossil record, its fossilised burrows a record of head-first, forward movement, and looking left, right, up and down for food. Then, in slightly younger rocks, just a little higher up the cliff, we can detect the sudden appearance of a host of new and recognisable and above all *different* animals – proto-arthropods, proto-vertebrates, annelid worms, penis worms, molluscs, arrow worms and more besides. This sudden burgeoning of animal diversity is famous as the 'Cambrian explosion', and it is an event in life's history that has obsessed palaeontologists and biologists since the early nineteenth century.

The real puzzle of the Cambrian is that these very different-looking animals seem to have no pre-existence recorded in the rocks that lie immediately below the Cambrian strata. It takes time for animals as different as arthropods and fish to evolve, which leaves us asking where they all suddenly appeared from. The mystery troubled Charles Darwin, who wrote in *On the Origin of Species*:

> if my theory be true, it is indisputable that before the lowest Silurian [Cambrian] stratum was deposited, long periods elapsed . . . and that during these vast, yet quite unknown, periods of time, the world swarmed with living creatures. To the question why we do not find records of these vast primordial periods, I can give no satisfactory answer.

What was it that lit the fuse on the explosion? What was the trigger that allowed the animals to blast out in these very different directions? What caused it to happen at this moment in time and no other? And, above all, could all this really have happened over such a short period – or was it more like the Cambrian smoulder? These questions, you may notice, are all about time. Explaining what caused the revolution means we need to know what else was happening in the world *at that moment*; to know whether it was an explosion or a smoulder, we have to know *how long* it really lasted. We need, in short, to be able to tell the evolutionary time.

The tree of life as we have thought about it so far has existed mostly in glorious isolation – a description of what was happening to different branches but with no reference to the outside world. A more complete history of life on earth requires context. What else was going on in the world when the first fish crawled onto land? Or when penguins abandoned flight? Or when the last of the giant coiled ammonites lay down and died at the bottom of an ancient ocean? What was the climate like when a human first cooked food, spoke words or drew a picture? Earth's 4.5 billion years have witnessed drastic shifts in the composition of the atmosphere; swings of global climate from snowball earth to greenhouse world; tectonic upheavals that have formed and broken supercontinents, moved these lands from equator to pole, pushed up mountain ranges, opened chasms between land masses and sprouted volcanic islands in the middle of oceans. And the evolution of every species is influenced by the species they happen to share the world with: the evolution of jaws in one branch of the animal tree may lead to the evolution of armour in another; the evolution of photosynthesis in cyanobacteria absolutely required other life forms to come up with ways to tolerate the presence of oxygen in the atmosphere and ultimately to exploit it. We need to include all these extrinsic forces as part of our story.

We could puzzle, for example, over the changing fortunes of the squat lobster-like trilobites, whose fossils are found in huge numbers in rocks deposited in the seas from the Cambrian period, roughly 520 million years ago, up to the end of the Permian, 270 million years later. Within the confines of the basic trilobite body, this group became hugely diverse, with thousands of different known species. Trilobite fossils range in size from something the size of a pinhead to a 72-centimetre-long monster;[2] some have elaborate eyes sticking out of their heads like a turret on a mediaeval castle; others have a huge head shield or long spines projecting from the body; one, *Walliserops trifurcatus*, has a spear almost as long as its body (and resembling Neptune's trident) sticking out of the front of its head.[3] What external forces promoted the early flourishing of the trilobites, the waxing and waning of their numbers, their changes in size and shape and, ultimately, their complete extinction?

There are some cases – though they are very rare – where the coincidence of a geological event and the fate of the organisms of the time is plain to see in the fossil record. The dinosaurs (along with a great many other species) were wiped out by a 6-mile-wide asteroid hitting the Gulf of Mexico, exploding with 10 billion times the power of the bomb dropped on Hiroshima and leaving a planet-wide sprinkling of iridium.[4] Such a monumental mass extinction event can be dated with unbelievable accuracy – recently, scientists were even able to establish the time of year the asteroid hit. This was done by studying fossil fishes that, from the impact debris found in their gills, we know must have died in the impact. Many fishes grow in fits and starts during the year, leaving recurrent marks in their bones like tree rings; and the unlucky fossil fish were found to have been approximately a quarter of the way through their annual growth cycle when they died. So it was springtime in the northern hemisphere, autumn in the southern, when the asteroid hit and the dinosaurs began to die, the asteroid's impact marking the abrupt start of a wave of extinction that saw off more than 70% of all species across the planet.[5]

But the fossil record is far from perfect, and such precise information is vanishingly rare. The chance of any species making a fossil is contingent on a perfect set of conditions, and this means that, while an individual from a species carrying a character we are interested in (the first fishes with legs perhaps) may well be lucky enough to be fossilised eventually, this fossil can only give us the very latest date for the appearance of this character. Fish legs might well have been around for millions of years before one of their owners was lucky enough to be both fossilised and (even more improbably) discovered by an interested palaeontologist. Dating events on the tree of life directly from the fossil record, in other words, is almost always hopelessly imprecise.

To tell the evolutionary time we really need not one but two accurate watches, the first giving the timing of the events in the geological record, the second telling us about branching points on the tree of life. Like bank robbers preparing for a heist, we need to synchronise these two independent watches. To calibrate our first watch, we must ask: how old is the earth? How old are different strata of rocks, and how old, therefore, are the fossils we find in these rocks?

The oldest recorded ideas about the age of the earth (and of subsequent events both mythical and historical) exist in religious accounts of creation, recorded, for example, in the Jewish Torah, the Egyptian Turin King List and the Zoroastrian Bundahishn. Most of these give a date for earth's creation of a few thousand to maybe a hundred thousand years before the present day. The exceptions to this rule are the Mayan and Hindu traditions, perhaps similar in their concepts of eternality and cycles of history. The Hindu units of time differ depending on the entity experiencing them, so the 100-year life of Brahma would correspond to over 300 trillion years for a mortal human.

The Western tradition for the date of creation, still accepted today by Young Earth creationists, is usually credited to Bishop Ussher, the fervently anti-papal Primate of Ireland and religious scholar (and far from the first or the last to calculate the years since the biblical creation).[6] Ussher is often derided for the precision of his calculation of the day (22 October 4004 BC), and even time of day (nightfall) of the biblical creation, but his was a brilliant feat of erudition needing a deep knowledge of ancient languages, history, astronomy and multiple, often contradictory, religious texts. His essential problem, given all the data, was to make an accurate join between Old Testament dates, counting forward from the creation until he could reach known historical dates, and then counting backwards from the present day to these same historical dates. The Old Testament dates begin, of course, with Genesis, and the next 2,082 years can be counted as those lived by Adam and Eve and their descendants (Seth, Enoch, Methuselah, etc.) up until significant events in the life of Abraham. From here, 430 years led to the Exodus from Egypt, and 480 years followed from the Exodus to the building of Solomon's temple. The reign of the Kings of Judah until the start of the captivity by Babylonians was then calculated at 424 years, and it is here that the join can be made with recorded historical dates – the death of Nebuchadnezzar and the enthronement of his successor, Amel/Awil/Evil Marduk. 'And it came to pass in the seven and thirtieth year of the captivity of Jehoiachin king of Judah, in the twelfth month, on the seven and twentieth day of the month, that Evilmerodach king of Babylon in the year that he began to reign did lift up the head of Jehoiachin king of Judah out of prison' (2 Kings 25:27).

That the earth could be older, even much older, than the biblical account of around 6,000 years became slowly evident in the discoveries of palaeontologists, geologists and physicists from the seventeenth century onwards. One of the most significant early pioneers of geology was the Dane Niels Stensen (or Nicolas Steno), who, alongside his contributions to science, converted from Lutheranism to Catholicism, became a bishop and, in 1988, was beatified by Pope John-Paul II.[7] In the autumn of 1666, as London was smouldering after the great fire, his patron, the science-loving Ferdinando II de' Medici, arranged for the head of a shark to be sent to Steno for study. As part of his dissection, Steno recognised that the teeth of the shark were essentially identical to unusual stones – known as *glossopetrae* or 'tongue stones' – that were commonly found buried in Tuscan rocks. Steno's intellectual leap (he was primed for this from previous discoveries of fossilised mollusc shells) was to understand that the *glossopetrae* really were shark's teeth, forcing him to ask how on earth teeth came to be found in the middle of a solid rock at the top of a mountain. His conclusion, extraordinary at the time, was that they had been placed there at the time of the rock's formation. The corollary was that the teeth, and the new layers of rock surrounding and surmounting them, must have been laid down on something already existing – an even older layer of rock – and that layers of rocks must therefore have been formed in succession, one above the other. Layered rocks, rather than being instantaneous creations of God, must have accumulated gradually through time.

Approximately contemporary with Steno, the English scientist Robert Hooke recognised the significance of fossils that resembled living marine molluscs but which were found buried in rocks far from the sea. He wrote that 'these will certify a Natural Antiquary, that such and such places have been under water, that there have been such and such kind of Animals, that there have been such and such preceding Alterations and Changes in the superficial Parts of the Earth'.[8] The rocks were a record, in other words, of the world as it was in the past, populated by unfamiliar animals that crawled across seabeds but were destined one day to find themselves at the top of a mountain.

Steno laid the foundations (appropriately) for the development in the nineteenth century of stratigraphy; he recognised that new rocks must be deposited upon an already existing substrate, and that these would naturally be formed in continuous horizontal layers under the influence of gravity. Deeper rocks must therefore be older than shallower ones, and non-horizontal rock layers – of which there are plenty to be seen – must have shifted over some long period of time from the horizontal. Rock layers, interpreted in this new framework, were hinting at processes in the past that must have lasted many, many years – time enough for vanishingly thin layers of soft sediment to build cliffs of solid rock a thousand metres high and for rivers to cut, millimetre by millimetre, deep valleys, canyons and gorges. The evidence that Steno and others were beginning to read in the rocks was incompatible with the idea of a young earth that had been carved solely by occasional catastrophes, such as the biblical great flood.

Calculating the true age of the earth was tackled in several ingenious ways. Darwin included a partial estimate in the first edition of *On the Origin of Species* in a section titled 'The denudation of the Weald'. The Weald is a shallow valley in Kent found between the North and South Downs (two parallel ridges of chalk). The downs are two sides of what was once a continuous dome of chalk, a thick layer of rock that formed in shallow

Cretaceous seas from the calcium carbonate in the shells of countless trillions of tiny marine organisms; the Weald itself is the space between the chalk ridges where the chalk dome has been eroded away.

Darwin, wanting to bolster his idea that the earth was old enough for the slow process of evolution to have happened, made a back-of-the-envelope calculation for the number of years that must have passed in order to erode this vast thickness of rock. The layer of chalk in the modern downs (and hence, once upon a time, the depth of the chalk in the dome) is 1,100 feet thick, and the breadth that would have had to be eroded – the distance between the North and South Downs – is 22 miles. Darwin guessed that a cliff of chalk 1,000 feet high would be eroded at the rate of half an inch per century and so, to have eroded the 1,393,920 inches required to produce the weald would have taken more than 360 million years. This was not terribly wrong – since the Cretaceous ended 66 million years ago, Darwin's estimate was only six times too old. And for Darwin's immediate purpose of revealing an old planet, it was certainly good enough.

The work that finally gave us the age of the earth and the solar system (and also of the individual layers of rock and of the fossils within them) came not from geologists, nor from biologists, but from physicists. An early attempt came from the eighteenth-century French cosmologist Georges-Louis Leclerc, Comte de Buffon. Buffon was a Plutonist (or Vulcanist) who believed (largely correctly) that the first solid rocks of the earth were formed from molten lava. This was in opposition to the Neptunists who believed the rocks of the earth to have been made by sedimentation in an ancient ocean. Buffon, from his Plutonist viewpoint, made the assumption that the early earth started off extremely hot.[9] With this in mind, he designed an experiment that involved heating a series of iron cannonballs of different diameters to a white heat to represent the hot, early earth. Buffon allowed these cannonballs to cool to room temperature, the larger ones cooling more slowly than the smaller. By plotting these measurements, he could then extrapolate to larger and larger spheres, ultimately calculating the time that a sphere the size of the earth would have taken to cool to its current temperature. He concluded that '22,595,086,068 minutes, which is to say forty-two thousand nine hundred sixty-four years and two hundred twenty-one days, is the time needed to cool a globe the size of the earth only to the point where you could touch it without burning yourself'. And '. . . ninety-six thousand six hundred seventy years and one hundred thirty-two days [is the time needed to reach] today's temperature'.[10] Buffon then adjusted these crazily precise numbers to account for earth being made not just of iron but also of other, more quickly cooling materials (he had made cooling measurements on a huge number of different materials) and came up with a final estimate for the age of the earth of 74,047 years – wildly wrong, but heading in the right direction.

Buffon had made an important mistake when he assumed he could extrapolate data on cannonballs (with a maximum diameter 13 centimetres) to the whole earth (diameter 12,742 kilometres). The error, pointed out by British mathematician and engineer William Thompson (later Lord Kelvin, the first scientist to be elevated to the House of Lords in 1892), was that, while a small cannonball effectively cools all at once, a larger sphere would maintain the heat in its core much longer, and this would keep the exterior warm over much longer periods through conduction.[11] The surface layer of a planet, in other words, would cool much more slowly than a tiny cannonball. Lord Kelvin's new estimate, in 1864, gave a much older (and much less precise) estimate for the age of the earth at between 20 million and 400 million years. Closer to reality, but apparently not nearly old enough for Darwin, who was caused to remove his 'denudation of the Weald' section from later editions of *On the Origin of Species.*

Lord Kelvin was wrong, and it is popularly believed that he erred because he knew nothing of the extra heat supplied by the radioactivity of nuclear fission. This, so it is said, led him to underestimate the quantity of heat being supplied from the centre of the earth, and so to underestimate how slowly the earth would cool – and hence the planet's age.[12] But the real source of Kelvin's underestimation was ignoring the additional transfer of heat by convection – rock below the earth's crust is molten and this hot moving liquid can continue to transport heat up to the surface keeping it warmer for far longer than would be possible by conduction through a solid planet. Lord Kelvin's former assistant, John Perry, ran his own calculations, giving an age of 2–3 billion years, which is really very close to the true 4.5-billion-year age of the earth.[13]*

Rather than being a source of confusion, nuclear fission turned out to be the clue that finally gave up an accurate age of the earth. Unstable heavy elements (uranium, plutonium, etc.) decay into lighter elements at an unwavering rate, measured in terms of their 'half-life', which is the time taken for half of the atoms of that particular heavy element to split apart and turn into the lighter. If we have a rock that, when it was formed, contained a known number of atoms of the heavy element, then the element's half-life tells us how many heavy atoms will be left after a given period of time. If we can count the number of heavy atoms remaining in the rock, we can therefore know its age. Isotopes of some elements, such as uranium-238 (U238) and uranium-235 (U235), have very long half-lives, allowing us to date very old events. Half of any quantity of U235 atoms will decay into lead-207 (Pb207) atoms over a period of 700 million years; the heavier isotope U238 decays to Pb206 even more slowly, with a half-life of 4.5 billion years.

Zircon is the most useful mineral for dating really old events. This is down to a piece of chemical luck. When zircon crystals form, they naturally exclude any lead atoms that might be around, but they are perfectly happy to include any passing uranium atoms. This serendipitous bit of chemistry means that we can be sure that all the lead atoms found in any present-day zircon crystal can only be the product of the radioactive decay of the uranium

* Poor Perry's hesitation is palpable in the first paragraphs of his letter to the journal *Nature*: 'I have sometimes been asked . . . to criticize Lord Kelvin's calculation of the probable age of the earth. I have usually said that it is hopeless to expect that Lord Kelvin should have made an error in calculation.'

that was originally there. A simple comparison of the ratio of uranium atoms to lead atoms tells us how many years ago a zircon crystal was formed. Finding a zircon crystal containing equal quantities of U235 and its decay product Pb207 tells us that half of the original uranium has decayed to lead. And we know from the unchanging half-life of U235 that this process must have taken 700 million years. From this simple measurement of the quantities of two elements we can accurately date the formation of this crystal to 700 million years ago; an older zircon would contain more lead and less uranium and vice versa.[14]

Radiometric dating has given us the means to complete the first half of our puzzle – we can find the absolute ages not only of the entire planet but also of the many different rock strata and of the fossils these rocks contain.

The rock in Newfoundland into which the golden spike was hammered has been dated to 538 million years ago. This gives us the date of the very first appearance in the fossil record of bilaterian animals like us, with a brain, mouth and eyes grouped together at the front, a gut that runs the length of the body, ending with an anus at the back, and, of course, left and right sides of the body that mirror each other.

The rocks have given us a date for the oldest fossilised complex animals we know about, but how do we know that these really are the first of their kind? The fossil record is well known to be gappy, which means that complex animals may be older than their first appearance as fossils. To test this possibility (and much more generally to reconcile the ages on the tree of life with the unique information to be discovered within the rocks), we have to set the time on the second watch. We need to find a way to measure the passing of time on the tree of life which will tell us the ages of our ancestors. The solution to this second problem requires us to find something in biology that is akin to the regular tick of a radioactive half-life. This biological pendulum was discovered hidden in something we now have in great abundance which is the steadily changing sequences of protein and DNA molecules.

15

Using Our Genes to Tell the Time

PART OF THE inspiration behind Ernst Haeckel's tree came from his University of Jena colleague August Schleicher, who, in the early 1850s, had started to think of languages as being things that could evolve, flourish and sometimes go extinct.[1] Schleicher, even before the publication of *On the Origin of Species*, proposed a tree model to describe the evolution of languages, but perhaps his grandest aim was to travel back in time, down the branches of his language tree, to reconstruct what he called the 'Ursprache' ('Ancient Language') – the common ancestor of all Indo-European tongues.[2]

A hundred years after Schleicher – using a method that bears the grand title of 'glottochronology' – the iconoclastic American lexicographer Morris Swadesh took this idea further when he attempted to estimate the date when different languages split apart (perhaps with the ultimate aim of dating the Ursprache). The way he attempted this is instructive because it was done in much the same way as ancestors on the tree of life are dated. Swadesh was an extraordinary linguist and a hugely energetic, moral and determined man.[3] As a linguist he had been provided with the perfect start in life, speaking English, Russian and Yiddish at home from his first years. At university he specialised in indigenous American and later Mexican languages as well as teaching himself, in the course of a year, to speak and to write Spanish well enough to give lectures, and to write the first of several textbooks in the language. He spent the later part of his career in Mexico having been branded a Red during the McCarthy era, for actions such as 'vigorously championing student demonstrators' and protesting against the death penalty for the spies Julius and Ethel Rosenberg. Swadesh was, indeed, a card-carrying communist.

As a part of his interest in the statistical analysis of languages, Swadesh proposed that when two languages separate (usually when two cultures become geographically separated) the words they initially shared (all of them) become replaced at a predictable rate over the span of centuries.[4] He drew up a set of 100 universal words ('all', 'and', 'animal', 'ashes', 'at', 'back', 'bad', 'bark', 'because', 'belly', 'berry', 'big', 'bird', and so on) that he expected to exist in all languages, regardless of any cultural, geographic or any other influences. Swadesh could measure how similar two languages are today by counting how many universal words they still share, but how could he transform this measure into a date of separation – how fast, in other words, should we expect languages to change?

To find the – hopefully universal – rate of change, Swadesh measured the number of differences between lots of pairs of languages with known divergence times – such as Middle Egyptian from Coptic, or Classical Latin from Romanian. He found that for the 100 words in his universal list, any pair of languages became different at a rate of about 25 words per 1,000 years. If his rule truly were to hold for all languages, this simple measure should allow us to date the age of the common ancestor of any pair of languages simply by counting how many of his universal words they still share. His method has proved to be less accurate than hoped thanks to the messiness of how languages actually evolve, but this example nevertheless shows us the path we might follow to find out how long ago any given ancestor on our tree of life existed. We might, for example, be interested in the age of the common ancestor of the bilaterian animals; fossils of these animals, you will remember, first appear in rocks that have been dated at 538 million years old, but might the first bilaterians be much older than this first step into the limelight?

To see how we might date this animal ancestor, let us first see how the genetic equivalent of Swadesh's list of words – the evolving sequences of biological molecules – might help. We have seen that the lengths of the branches on evolutionary trees are intended to represent the amount of change that has happened, so that the branch separating a human from a chimpanzee – for example – is relatively short because they are physically (phenotypically) similar and the sequences of nucleotides in their genes are similar too, whereas the branch separating a human and a jellyfish is much longer, representing lots of morphological and genetic differences. Just as Swadesh showed us, if we measure how different two species are, as long as the rate of change is predictable, we should be able to calculate how distantly related (in years) they must be, that is, how long ago they had a common ancestor.

While we should certainly expect there to be a rough relationship between time and evolutionary change, a clock is only going to be useful if it is consistent. To see the sort of problem we might find on a human timescale, think about using a child's height to tell how old they are – height will of course reliably increase as they get older, but it is far from being an accurate measure of their age: some kids are destined to be tall and some short; some adolescents will shoot up early and some (me for example) will be late bloomers. We need the rate of change to be predictable over huge periods of evolutionary time, just as a ticking clock fills hours, days, years with a definite number of seconds.

We have two types of data whose rate of change we might use to measure the passage of time: morphological change and genetic change. We have already seen hints that morphological change might not fit the bill as a reliable pendulum. We have met living fossils like the coelacanth, whose form, cocooned in its stable environment, has barely changed since the Early Devonian (410 million years ago); over the exact same time period, some of the coelacanth's relatives have evolved into frogs and bats and snakes. Coelacanths, frogs, bats and snakes have evolved for the exact same number of years, starting from their lobe-finned common ancestor, but their degrees of morphological change are utterly different. Adding up morphological changes cannot possibly reliably tell the time.

If we wish to find a regularly swinging biological pendulum, we will need to look for characters whose rate of change is much less influenced by the environment and the effects of natural selection. We need to look at the level of our genes.

The realisation that sequences of amino acids in proteins and of nucleotides in genes could demonstrate the steady change we are looking for came shockingly hard on the heels of the very first sequences of proteins being painstakingly read in the late 1950s. The first version of what we now call the 'molecular clock' appeared in 1962 in a long paper that is more broadly concerned with what the new sequences of proteins might tell us about disease.[5] The paper, written by Emile Zuckerkandl and the double Nobel Prize-winning (Chemistry and Peace) Linus Pauling, includes an extended section on how protein sequences evolve. They focus on haemoglobin ('hemoglobin' in America), the protein found in vertebrate red blood cells that carries oxygen, and for which several amino acid sequences were then available. Zuckerkandl and Pauling noted that as two species grow to be more different, so the genes that reside within them should change in tandem. The evidence for this tendency could be found in the haemoglobin sequences of various vertebrates that became increasingly different from human haemoglobin as more distant relatives were compared.

In other words, more distantly related species have more different proteins. So far so expected. Then they say, with zero fanfare for an insight that will have such an impact, 'It is possible to evaluate very roughly and tentatively the time that has elapsed since any two of the hemoglobin chains present in a given species . . . diverged from a common . . . ancestor.'* They go

* They are referring here to dating the divergence between the duplicated genes within a species – adult and foetal haemoglobins for example – but the same principle applies to dating divergence between species.

on to make the very first attempt to use this method to date an actual ancestor – that of humans and gorillas – and we will follow their account, which shows the three steps we must take when using molecular clocks.

The first thing they do, sensibly enough, is count the number of amino acids that differ in the haemoglobins of several different mammals; they note in particular that there are eighteen differences in amino acid sequences between the haemoglobins of horse and human and just one or two differences between gorilla and human. Their second step is to acknowledge the importance of change happening regularly over time: 'it is assumed . . . that the evolutionarily effective mutation rate . . . fluctuated . . . around a mean without showing a predominant trend to increase or to decrease'. That the number of mutations per million years is fairly constant.

The third and final step is to use some prior knowledge to calibrate their clock – to know how fast it ticks. They need to work out just how many million years a single mutation corresponds to. This is normally the point where the molecular clock meets the fossil record, because, to calibrate any molecular clock, we need evidence from at least some part of the tree where we know for sure the age of an ancestor.

The calibration of the clock is essential. Knowing that there is a 10% difference in the sequence of a certain gene in two species doesn't tell you how long ago their common ancestor lived unless we also know that that gene changes at, say, 1% per million years. This is the same as how knowing the number of miles you have travelled can tell you how long ago you left home but only if you also know how fast you were driving. For molecules, the rate can be measured as percentage change per million years, and to get the right figure, we need to know the age of at least one ancestor on our tree. Zuckerkandl and Pauling rather pull a rabbit out of a hat by making the (incorrect) assumption that the common ancestor of humans and horses lived 'in the Cretaceous or possibly in the Jurassic period, say between 100 and 160 millions of years ago'. Eighteen changes of amino acid in 100 million to 160 million years means the clock very roughly ticks once (on each branch) every 7 million years.

Calibrating molecular clocks is never as straightforward as Zuckerkandl and Pauling's finger-in-the-air estimate might suggest (not least because the common ancestor of humans and horses actually lived about 85 million years ago). Part of the problem comes from the patchiness of the fossil record, which, as we have seen, means it is almost impossible to find a fossil corresponding to the moment a branch on the tree appears. The other problem of calibrating our tree with fossils might be obvious if you try to imagine a fossil that is the common ancestor of a human and a horse. What on earth would this look like? How would we recognise it if we dug it up? Despite the many difficulties, clever palaeontologists have been able to give confident dates for many different branches on the tree of life, and these act to anchor the tree in time. We can use these known points to calibrate the tick rate of the molecular clock across the whole tree.

With their clock optimistically calibrated at one change per branch per 7 million years by reference to the date of a 'known' ancestor, Zuckerkandl and Pauling went on to estimate the age of the extinct ape that was the common ancestor of humans and gorillas (which, remember, differ by only one or two amino acids in their haemoglobins*). In the 1960s, biologists generally believed the human-gorilla ancestor to have lived as much as 35 million years ago. Zuckerkandl and Pauling estimated that the human-gorilla ancestor lived much more recently, roughly 11 million years ago. Their estimate, a pure fluke given the imprecision of their clock, was nearly right. Today we believe this ancestor to have lived about 10 million years ago.

The next advances followed quickly. While the regularity of the molecular clock's tick was an article of faith for Zuckerkandl and Pauling, in 1963 the Egyptian-born biochemist, Emanuel

* The uncertainty – one or two differences – comes from our having more than one haemoglobin gene.

Margoliash, working in Chicago,[6] provided evidence to demonstrate this assumption was fair.[7] His way of testing it began with the observation that a bird is evolutionarily distant from all mammals to exactly the same degree.

Take the chicken. From the starting point of the common ancestor of birds and mammals, a present-day chicken will have evolved for 310 million years. Equally, every mammal species alive today must be separated from this common ancestor by the exact same period of time. The time that separates the chicken from *every single mammal alive today* is therefore identical (2 × 310 million years). Knowing this is hugely useful because it implies that if (and only if) the molecular clock ticks at a constant rate, the number of amino acid differences between the proteins of a chicken and each of the living mammals should be approximately equal. In other words, as long as the molecular clock ticks regularly, we should expect the same number of differences between a chicken and a human as between a chicken and a dog, a chicken and a dolphin and a chicken and a zebra.

Margoliash's results agreed pretty closely with this idea. Working with the amino acid sequences of Cytochrome C proteins (which have a role in the cell's mitochondrial power station), he compared the number of differences between a chicken and a pig, a rabbit, a human and a horse. The number of differences he found are almost identical with 12, 10, 11 and 14 changes. And going deeper in the tree: between a tuna on one branch and a chicken, pig, rabbit, human and horse on the other, he found 19, 17, 19, 21 and 18 differences.* Margoliash had effectively shown that the molecular clock (unlike the morphological clock) had ticked at a fairly constant rate over hundreds of millions of years.

* Margoliash, especially in collaboration with Walter Fitch, went on to be one of the pioneers of using molecular sequences to make evolutionary trees, starting with the Cytochrome C sequences he had produced in his lab. His lab's data eventually included the first molecular sequences from a coelacanth. As President de Gaulle had designated the fish a French national treasure (in 1970 the Comoro Islands where it was caught was a French colony), Margoliash was obliged to travel to France to carry out his experiments.

Using a molecular clock to date major events and the existence of key ancestors on our tree of life, and lining them up with the geological record (dated using radioactive half-lives), is a very exciting prospect. It is like adding an essential instrument to our time machine – a depth gauge, perhaps, that tells us just how far down into deep time we have travelled. While the idea behind molecular clocks is now sixty years old, its use still provokes passionate disagreement. One thing that has become clear is that, however isolated from the coalface of natural selection genes may be, the molecular clock is not *entirely* regular. Genes on different branches of the tree of life have evolved at different rates. More generally, there are so many little sources of uncertainty over how they tick that, if we take all of these into account, any estimates we come up with for the date of a long-extinct ancestor can be so broad as to be of little use. If we want to know whether the major groups of mammals appeared immediately after the extinction of the dinosaurs 65 million years ago, it is of little use to have an estimate for the mammal's appearance that is 'somewhere between 40 and 90 million years ago'.

Furthermore, molecular clock estimates sometimes give dates for past events that seem to fly in the face of the fossil evidence. The most common problem we meet is when the molecular clock tells us that a big branch of the tree seems to have existed for tens or even hundreds of millions of years before leaving a fossil as evidence of its existence.

These problems have some palaeontologists up in arms, none less energetic than my friend Graham Budd. Graham is distinguished by his amusing and idiosyncratic manner of speaking (the way he talks acts as a kind of meme, spreading itself by infecting the speech patterns of those around him), by the stick he walks with (named Mr Stick) and by his unusually broad and deep knowledge of the scientific literature. He is also a talented amateur pianist whose greatest boast is that he once accompanied Nobel Prize-winner Christiane Nüsslein-Volhard singing Schubert lieder. Graham is the academic son (I mean PhD student of) the

Cambridge palaeontologist Simon Conway Morris and therefore the academic grandson of Simon's PhD supervisor, Harry Whittington. Conway Morris and Whittington were both actors in the *Anomalocaris* story, you may recall, and both are rightly famous for their redescription and reinterpretation of the famous set of ancient animal fossils called the Burgess Shale fauna. These are a beautifully preserved example of the complex animals whose fossils made their first sudden appearance in the rocks above the golden spike in Newfoundland. One of the most fiercely fought battles over the use of the molecular clock concerns the true age of the ancestor of these Burgess Shale animals.

The Burgess Shale fossils were discovered in 1909, high on the side of a mountain in the Canadian Rockies close to known trilobite fossil-bearing rocks. The chance discovery of the Burgess Shale itself was made by the Canadian geologist Charles Doolittle Walcott. The story goes that he discovered the marvellous outcrop when his wife Helena's horse disturbed one of the fossil-bearing rocks (it contained a dainty crab-like fossil called *Marrella*) that had slid down the mountainside to lie by the trail. This tale, reported in his obituaries, seems to be an enjoyable, often repeated and frequently mangled origin story that probably has only a loose relationship with the truth. What is sure is that the rock formation that the *Marrella* slab had detached from was discovered the following year, and Walcott, his wife and children and various other collaborators spent many more summer seasons collecting there.[8]

In the few hundred million years leading up to the Cambrian, a series of glaciations had affected huge regions of the globe, bringing ice almost to the equator; and sea levels had been very much lowered thanks to all the water getting locked up in ice. As the climate warmed in the Cambrian, the ice melted, and the rising oceans washed up over the edges of the continents to produce warm, shallow seas. The Burgess Shale animals, which date from the Middle Cambrian, roughly 508 million years ago, lived on the edge of a continent called Laurentia which at that time was on the equator.

The Burgess Shale animals are the standout stars of Cambrian fossils because of the way that they were preserved. The animals that produced the fossils lived in shallow waters, perched on a ledge at the top of an underwater cliff. As portions of the cliff occasionally broke off and sank to the depths of the ocean, they would take a handful of these animals with them. The lack of oxygen in the deep waters was the magic that prevented the dead animals from decaying immediately (or from being eaten by passing scavengers), and they were quickly covered by a sprinkling of fine sediment. These conditions resulted in the preservation of wonderful details of their morphology, including the soft parts, which normally rot quickly away leaving just bones, teeth and shells. The Burgess Shale is a so-called *Lagerstätte*, a trove of spectacularly preserved fossils.

The sudden appearance of bilaterian animals (and later of the Burgess Shale fossils) is strange enough, but the mystery considerably deepened when the molecular clock was used to calculate the age of their common ancestor. These first studies estimated that the bilaterians first existed a billion years ago – perhaps even longer.[9] The very oldest bilaterian fossils are less than half this age. Thanks to better methods and data, the gap between the molecular clock estimate for the age of the first bilaterally symmetrical animals and their first appearance as fossils has shrunk to a more reasonable 120 million years.[10] But we need to put this fossil-free stretch of the Precambrian in context: it is a stretch of time longer than the entire 100-million-year history of the placental mammals, in which not a single specimen from these early animal groups was preserved in a fossil. Where on earth, we might ask, were they hiding?

If bilaterians really existed for 120 million years without leaving a single fossil, then we need an explanation. It may simply be that we have not looked in the right places, or that they lived in regions that are now inaccessible to the modern palaeontologist. But the more popular explanation invokes the biology of the Precambrian animals themselves. It says that if bilaterians were around long before the Cambrian explosion, they must have been tiny – microscopic even. The sudden appearance of animals in Cambrian rocks would therefore represent not the first moments of their existence but rather the global impact of some external force – increasing oxygen levels, perhaps – that triggered a sudden increase in their size. According to this idea, animals didn't suddenly come into existence, they just became big enough to leave fossils.

To see whether this story of tiny Precambrian animals makes sense, we can ask whether the living descendants of the very first branches of the bilaterians are tiny and simple – unlikely to make a fossil and easy to miss even if they do. And this question takes us right back to the animal tree of life, where we must first ask: which *are* the very first branches of the bilaterians?

I have led you down a convoluted path from geological and molecular clocks to the age of an ancient animal ancestor, to a puzzle of some missing fossils, and we have ended up right back with the animal tree of life. The question about the first branches of the bilaterian branch turns out to concern some of the smallest and simplest and least known of the animals; it is, therefore, more than a little esoteric, but it is the cause of an argument that I have been deeply involved in for more than two decades.

The problem boils down to working out the true place on the tree of life of a group of (mostly) tiny, very simple, easily overlooked worms – these are the worms that many people believe are the earliest branch of the bilaterians we are looking for (and whose tininess and simplicity would neatly explain the lack of Precambrian fossils). The worms are called the acoelomorphs, and we have met them before in the mint sauce worm *Symsagittifera roscoffensis*.

I am undoubtedly swimming against the tide, but I belong to a rather small group that disagrees with the idea that these worms are the tiny and simple first branch of the bilaterians. My own view is that the acoelomorphs, along with another equally humble relative of theirs called *Xenoturbella bocki*, are related to some large and complex animals, including the vertebrates. If I am right then these simple worms tell us nothing about the size of the first bilaterians and do not help solve the mystery of the missing Precambrian animals. The story of the wanderings of the acoelomorphs and *Xenoturbella* around the animal tree of life, and the arguments they have provoked and continue to provoke, will be told in the next chapters, where I will indulge myself by talking about these and some other lesser-known worms that have been my passion for more than thirty years.

16

Embryos and Arrow Worms

If you were to scoop up a bucket of water in the English Channel in the late summer months and place a droplet under a microscope, you would almost certainly see, floating amongst a host of tiny planktonic crustaceans and larval fish, a single, crystal-clear cell sitting inside the sphere of an equally clear membrane. It will be large for a cell, perhaps a millimetre across, and the outer membrane tough but giving under the prod of a glass pipette. I am going to show you how you might discover a lot about this cell – especially its identity – simply by waiting and watching.

The size of the cell, many times bigger than a bacterium, tells us immediately that it belongs to the great branch of eukaryotes. This means there is a strong possibility that it is a single-celled organism and therefore will not do very much at all. But things get a little more interesting when our cell soon divides, and we see that the two daughter cells remain stuck together; these two daughters themselves divide again and again, building up into a hollow sphere of cells all copied from the original. Our patience has been rewarded with a big clue as to what this species is: we can, in this game of *Guess Who*, rule out all the single-celled eukaryotes. We have narrowed it down to one of the small number of eukaryotic branches containing multicellular species. These multicellular eukaryotes comprise the three kinds of algae (red, green and brown), the green plants, some fungi and all the animals. Our cell has no chloroplasts, so we can rule out algae and plants, and marine fungi never look this complex, so we are left to conclude that our growing organism must be an animal. But what kind of animal?

What happens next – the complicated movements of the cells during the rest of embryonic development – can give us more clues as to which of the main branches of the animal tree of life this little embryonic animal belongs. An indentation will appear and penetrate into the hollow ball to form a tube (its future intestine); later on, blocks of muscle and a nervous system will appear; then eyes, fins and a tail; then a mouth surrounded by fierce spiny jaws. Our beautiful transparent cell is going to morph at the end of all of this into a marine worm called an arrow worm, whose uncertain place on the animal tree of life has been argued over for more than 150 years. For most of this time, these many elements of its embryonic development were central to the debate.

One important late-nineteenth-century innovation in working out the relationships between distantly related animal groups came in large part from Ernst Haeckel. He suggested that the way that animals are built during embryonic development can give clues about relationships between creatures whose fully grown forms seem irreconcilably different. We have seen how the tadpole larvae of the sea squirt (first thought by Haeckel himself to be related to molluscs) showed that their real affinities lay with the vertebrates. Embryological similarities have also been used to discover, for example, that barnacles (which had been thought to be related to molluscs, like the limpets they live beside) are really crustaceans (close relatives of crabs and lobsters).

The Austrian zoologist Karl Grobben was one keen adopter of Haeckel's ideas. Grobben could so easily have been forgotten today, eighty years after his death (the inevitable fate of almost all scientists). Most of the key events of his sparsely documented life and solid career can be quickly listed: he was born in 1854 in Austrian Brno; he took his undergraduate and PhD degrees in Vienna; he made some important (but now forgotten) discoveries concerning the embryology of crustaceans and molluscs; he had half a dozen species named after him (including an endangered gerbil); and he edited a new edition of a famous zoology textbook (the original written by his mentor Carl Claus).[1] At university in Vienna, Grobben was a fellow student of Sigmund Freud – the two of them won scholarships to the zoological station in Trieste where Freud worked on the sexual organs of the eel (and also started the smoking habit that would eventually kill him).[2] In another brush with greatness, Grobben's brother-in-law, Erich von Tschermak, was one of the rediscoverers of Mendel's work on the genetics of peas. Grobben was undoubtedly a successful academic – full professor at the age of thirty-nine, director of an institute and elected a fellow of various august academies – but the fact that his work continues to have an impact today is down to a paper he wrote in 1908.

Grobben's 1908 paper 'Die systematische Einteilung des Tierreichs' ('The systematic division of the animal kingdom') outlined his views on the relationships between the most distinct animal groups.[3] Using a select few of the various characteristics of animal embryology, he divided Bilateria into two great branches that he named the protostomes and the deuterostomes.

The difference Grobben noted between the two branches comes down to that early event in an animal's development when the embryo is a hollow sphere of cells, and the indentation that will form the gut begins working its way inside (imagine pushing your fist into a balloon to form a tunnel). The initial opening of this gut-tube to the outside world (the place where your arm would be pushing into the balloon) has two possible fates in different animal groups. In one of Grobben's great branches, the protostomes, the opening of the tube is destined to become the mouth end of the gut (*proto stoma* meaning 'first mouth'). In the other branch, the deuterostomes, the opening becomes the anus, and the mouth opening, at the other end of the gut, will form later (*deutero stoma* meaning 'second mouth').

Grobben's protostome branch includes almost all familiar big groups of invertebrate animals, including the arthropods, nematode (round) worms, annelid worms, molluscs and flatworms, as well as at least a dozen other much less familiar phyla. His deuterostome branch is smaller – it contains the echinoderms ('spiny skins' which are familiar as starfish and sea urchins), the chaetognaths (arrow worms), a little-known group called the hemichordates and the chordates – most important of which are the vertebrates.* You will probably agree that starfish (echinoderms) really don't seem an obvious choice for close relatives of humans (vertebrates). If Grobben had correctly established their kinship solely by comparing their embryological development, that would certainly vindicate Haeckel's ideas about the usefulness of embryonic characteristics.

My own interest in Grobben's paper began when I was researching the chaetognaths (arrow worms) for my PhD. My aim was to test Grobben's idea that (along with starfish) the arrow worms were deuterostomes and so one of the very closest invertebrate relatives to our branch of vertebrates. When I began my work, Grobben's idea about where arrow worms fit in the animal kingdom had endured for more than eighty years; in any zoology textbook from the twentieth century, the arrow worms can reliably be found in the penultimate chapter just before you get to the vertebrates. If I could prove this close relationship between arrow worms and vertebrates using information in DNA sequences for the very first time, we could be confident that studying arrow worms would provide valuable clues about what the ancestors of the very first vertebrates looked like.

Adult arrow worms are at the same time beautiful and horrifying. They are long and slender and crystal-clear, like tiny slivers of broken glass. But they are voracious predators with terrifying jaws – terrifying at least if you are one of the tiny crustaceans or fish larvae they eat in huge quantities. They also have absolutely gorgeous embryos, about a millimetre across and completely transparent, as we have seen. Their transparency makes it unusually easy to follow their embryonic development. And as we follow the proliferation of their embryonic cells, the grouping of these cells into a hollow ball, the indentation pushing into these cells and so on, what we see is that arrow worms do indeed turn the opening to their embryonic gut not into a

* Sea squirts are chordates, as are another simple fish-like animal known as the lancelet or amphioxus ('sharp at both ends').

mouth (like the protostomes) but into an anus just like the starfish and vertebrates.

My PhD experiments would compare a snippet of arrow worm DNA with the same stretch of DNA from a handful of other animals. In those days, sequencing a single gene from half a dozen animals was difficult and time-consuming – in fact, probably half of the thirty-odd animal phyla had not yet had any of their DNA sequenced. The only sensible choice of gene to study at that time was the small subunit ribosomal RNA I have already talked so much about. Cutting a longish story short, what I discovered was that Grobben's widely believed idea for the relationships between arrow worms, starfish and vertebrates (based on embryology) was wrong. The comparison of SSU rRNA genes showed that the arrow worms are more closely related to flies and snails (protostomes) than they are to vertebrates and echinoderms (deuterostomes). The similarities of arrow worm, vertebrate and starfish embryology had proved to be a red herring – and now required an explanation.

Correcting this little bit of the tree of life was an achievement, of course, and has some intrinsic importance simply because we want to know the shape of the tree of life. But the real pay-off was that a part of the story of animal evolution had changed. I had started with a theory in which both arrow worms and vertebrates emerged from a *relatively* recent common ancestor. If this relationship were true, our parsimony method would tell us that their common ancestor must have shared their unusual way of making an embryo, and that this had been passed on to both vertebrates and arrow worms. My discovery of a more distant relationship required a new explanation of their similarities. We are left with two interesting possibilities: either vertebrates and arrow worms both coincidentally evolved their unusual embryology through convergent evolution; or this way of making an embryo is a very ancient one that has been altered or lost in almost all animals except, for some reason, vertebrates, starfish and arrow worms.

Looking back at the paper I eventually wrote in 1993 with my doctoral supervisor Professor Peter Holland (now Linacre Professor of Zoology in Oxford), I can see that our results were not as clear as I had remembered; we made some mistakes, some of the results contradict each other and the evidence was rather weak.[4] We correctly concluded that arrow worms are not deuterostomes, but I am not sure we convincingly showed where they did fit. Our conclusion was nevertheless soon backed up by other research,[5] and this new place for the arrow worms in the animal kingdom – with the protostomes – was quickly accepted by the wider community.

The question of the closest invertebrate relatives of the vertebrates seemed to be settled once again. The vertebrate branch is closest to the echinoderms (and the rarely encountered hemichordates), and all the other bilaterian phyla, including the arrow worms, belong to the protostomes. I would think little more about this question until, ten years later, I became involved ('embroiled' might be more accurate) in a debate about where a second group of marine worms belongs, and whether these lowly worms might contain the clues to vertebrate origins that seemed to have been lost when we removed the arrow worms from the deuterostome branch.

17
The Diet of Worms

THE MARINE WORM *Xenoturbella bocki*, which passes its life in the mud at the bottom of a Swedish fjord, seems an unlikely candidate for a close relative of our own noble branch of vertebrates – the zoological equivalent of an embarrassing uncle at a family party. However, while not everyone agrees, my own work over two decades, together with various members of my lab and several collaborators, suggests that *Xenoturbella*'s current reduced circumstances hide a more exalted former existence as a member of the deuterostome family – a relative of starfish and vertebrates that has fallen on hard times.

Almost every specimen of *Xenoturbella bocki* ever seen by humankind has been collected in the same way and in roughly the same location: in the waters of Gullmarsfjorden (*Gullmarn* means 'God's Sea' in Old Norse). The fjord sits about 70 kilometres north of Gothenburg on the west coast of Sweden and cuts about 25 kilometres inland from the Skagerrak, but it can be crossed close to its mouth by a pair of small car ferries painted daffodil yellow and named *Gullmaj* and *Gullbritt* (after, I like to imagine, the Swedish actress Maj-Britt Nilsson). Kristineberg marine station is found just inside the mouth of the fjord on the south shore, a short walk from the little town of Fiskebäckskil, which is a very lovely place, filled with clapboard summer houses and a few chi-chi (and pricey) restaurants on the water's edge overlooking the yachts in the harbour. Kristineberg and Fiskebäckskil are especially wonderful in the long Swedish summer evenings. The winters – I have been once in November – are a different matter. The marine station is one of the oldest in the world, established by the Royal Swedish Academy of Sciences in 1877, and for over two decades has provided us with lab space and the use of a small research ship called the *Oscar von Sydow* along with the expertise of the ship's captain plus a crew of one.

To collect *Xenoturbella*, the first thing you need to do is to sail to exactly the right spot on the fjord. The entire fjord covers an area of almost 100 square kilometres, but the most reliable collecting spot has proved to be a single football-pitch-sized patch on the north side of the fjord opposite the marine station and near the town of Lyseskil. The worms live in the top layer of the mud at the bottom of the fjord, about 50 metres below the surface. To collect them we lower a device called a Warén dredge on the end of a long wire rope; this aluminium sledge is slowly dragged along by the boat, skimming off the top layer of sediment from the bottom of the fjord until it is filled with mud. The full dredge is then winched up and the mud is (very messily) emptied into big plastic boxes. We collect a couple of hundred litres of this sloppy mud in three or four large plastic boxes to take back to the marine station for the next step.

To find, if we're lucky, half a dozen worms in all this thick brown sludge, we need to get rid of all the mud; this is a very fine silt and easily washed away. We do this by scooping the mud into the first of a pair of sieves, stacked one above the other; this has large holes and only captures big objects – usually there are worm burrows made of solidified mud, heart-shaped 'irregular' sea urchins and occasionally a fish. Smaller things pass into the second, smaller-holed sieve, and once all the fine mud has been washed away, we are left with a lot of small debris including small annelid worms, bits of broken shell, some tiny bivalve molluscs and, hopefully, hidden away in a chaos of other bits, some live *Xenoturbella* specimens. Finding the worms amongst the detritus is an art that comes with practice but which always involves hours of staring at the contents of a slightly smelly plastic tray. It is a tedium enlivened by occasional flashes of excitement at finding what often turns out to be an unwanted flatworm.

I began thinking about *Xenoturbella* in 2002 when I was at

Cambridge working in the Department of Zoology. Our work on *Xenoturbella* was at first exciting and surprising but after a while became controversial; an argument has developed that I have been enjoying (when I am winning) and enduring (when my opponents seem to have the upper hand) ever since. I care a lot about this; I am still doing occasional research on it, and the argument continues to provoke the occasional anxiety dream. It is not easy to explain why I should care so much, but I'm going to try.

A big part of the reason it is hard to explain is the unimpressive-looking protagonist: not a huge dinosaur, a stunning peacock or a charismatic lemur (or even a lovely arrow worm) but a small, brownish worm that leads an uneventful life browsing the mud for small molluscs to eat. While we know what it eats (and its diet is the MacGuffin of an early plot twist), we know little more about what else it might get up to because the only specimens that have ever been seen have been dragged up from their natural home.

Xenoturbella was discovered in Gullmarsfjorden in 1915 by Karl Alfred Sixten Bock. But it wasn't until 1950, four years after Bock's death, that another Swedish scientist, Einar Westblad, published a paper that described *Xenoturbella*, gave it a name (the species name of *bocki* recognised the work of Bock) and puzzled over its place in the tree of life.[1]

From the perspective of someone with an interest in animal evolution at the broadest scale, the most important facet of *Xenoturbella*'s biology is that it really is a very simple animal. It has no real brain (its nerve cells are spread around its whole body) or organs of any kind, it has no cavities within its body (coeloms), and, perhaps weirdest of all, it has only one opening to its gut – it has no separate mouth and anus, meaning that food goes in and waste comes out of the same hole. All of these absences make it look simple (some might say primitive), and you might sensibly conclude that *Xenoturbella* must be a survivor of an early branch of animals that separated from things like vertebrates, molluscs and arthropods (which have all of the features that *Xenoturbella* lacks) before these advanced features evolved.

There is a second – almost equally – simple phylum of animals called the flatworms (Platyhelminths) whose perceived relationship to the other animals is instructive. The best known of the flatworms are nasty parasites like tapeworms, liver flukes and schistosomes (which cause bilharzia), but they also include some often lovely, beautifully patterned, non-parasitic species known collectively as turbellarians. For much of the twentieth century, the simplicity of the flatworms (which, like *Xenoturbella*, have no separate mouth and anus and no body cavities/coeloms) was taken to indicate that they belonged to the earliest branch of the animals.

The first DNA sequences from flatworms showed, to the surprise of most zoologists, that the simple flatworms are related to complex animals like molluscs and annelid worms. Their true place on the tree of life tells us that their simplicity is misleading; flatworms, like sea squirts, orthonectids, dicyemids and various other simple animals we have met, started out their evolutionary journey as complex beasts. At some point in their history, and for reasons we may never know, they have lost some of this complexity (it's hard to fathom why they have abandoned a separate mouth and anus!).

Returning to *Xenoturbella*, Westblad obviously thought the most likely branch to accommodate it was the one that leads to the simple flatworms (Turbellaria). His choice of prefix, however (*xeno* meaning 'strange'), conveys the idea that *Xenoturbella* looks even simpler, even more primitive than a typical flatworm, lacking even more of the features seen in most other animals. Westblad's opinion on *Xenoturbella*'s place on the tree of life is obvious from the title of the paper he wrote describing and naming the worm: '*Xenoturbella bocki* n.g., n.sp., a peculiar, primitive Turbellarian type'.* But before we get to the argument over where *Xenoturbella* really belongs, we have to follow an unexpected path that leads to a slightly awkward dead end.

In 1995, the Danish zoologist Claus Nielsen published a wonderful book outlining in glorious detail his ideas on the relationships

* n.g. = new genus; n.sp = new species

between the different animal phyla, and the characteristics he relied on to distinguish different branches of the animal tree. Nielsen ended his book with a chapter titled 'Five Enigmatic Taxa', where he pondered over groups of animals that he still had difficulty placing. One of these enigmas was *Xenoturbella*. But fast forward six years to 2001 and the second edition of Nielsen's book, and we find that he no longer included *Xenoturbella* amongst his enigmas. In truth, far from a mystery solved, *Xenoturbella* had been launched into a new era of confusion and even minor fame.

Nielsen's belief that the mystery of *Xenoturbella*'s place on the tree of life was solved came from two papers that had been published in 1997 in *Nature*.[2] Both papers, independently, came to the same surprising conclusion that *Xenoturbella* was a strange-looking mollusc. More than this, they claimed to show that it belonged not just somewhere close to the very diverse phylum of molluscs but as a member of a particular group, the protobranchs, that sits bang in the middle of the bivalve molluscs (oysters, mussels, clams, etc.).

The papers each came to this conclusion using very different methods. One had studied old specimens of *Xenoturbella* that had been sectioned (i.e. sliced up very thinly so it was possible to see right inside) by Bock and by Westblad early in the twentieth century. The author claimed to have found embryos buried within the skin of the adult worm that looked just like mollusc larvae. The extraordinary interpretation was that *Xenoturbella* embryos, growing in the epidermis of their parent, pass through a stage that looks just like an oyster larva before metamorphosing into an adult that definitely does not look like an oyster. This is certainly an unusual idea, but then a human baby grows right inside its mother's body; and the human embryo passes through a stage with fish-like gill buds and a tail before changing into something tailless and very un-fish-like. The second paper, more simply, reported that the first DNA sequences to be extracted from a *Xenoturbella* specimen closely resembled the equivalent sequences from bivalve molluscs.

Either one of these odd results might have raised a few eyebrows, but the mutual support they gave each other meant that they were widely accepted by the scientific community. One of the authors, in what was (as we will discover) a salutary example of hubris, reported 'the previously unknown embryology of *Xenoturbella* that unequivocally corroborates a bivalve relationship and thus once and for all dismisses the potential new phylum.' He goes on to recognise the strangeness of the result: 'Why would an animal that is neither parasitic nor microscopic nor short-lived lose all its organs and change its concentrated nervous system?'[3] Good question.

I find it impossible to reconstruct how I came to doubt these results. I still have boxes of old 3.5-inch floppy discs that must contain my emails from the early 2000s, but I am not quite motivated enough to find a way to read them to remind myself how things came about. What I remember is meeting Claus Nielsen in Amsterdam, perhaps in 2001, when we were both examiners of a PhD thesis. He gave me a lift in his little red car, and I brought up my suspicions about the papers linking *Xenoturbella* to the molluscs. My scepticism probably came from the unlikely seeming link to the bivalves, but there was also the deadly fact that at least part of the diet of *Xenoturbella* was known to be bivalve molluscs. While *Xenoturbella*'s diet couldn't really explain the mollusc-like embryos buried in *Xenoturbella*'s skin, it might at least explain the mollusc DNA that had been found in the tissues of an adult animal.

Nielsen, who died in 2024 after a long and distinguished career, was a fantastic field zoologist who spent his professional life not in a molecular biology lab or sitting in front of a computer analysing DNA sequences, but travelling the world observing, collecting and identifying marine animals.[4] Nielsen was familiar with *Xenoturbella*, knowing where and how to collect it, and he agreed to help us gather some specimens ('us' by this stage included Dr Sarah Bourlat, who had joined my lab in Cambridge).

We decided to compare the DNA sequences from two *Xenoturbella* specimens: one whole, including gut, last meal and all; and a second from which we had removed the intestine (and with this any possible source of contamination from its food). First, Bourlat used various chemicals and enzymes to purify the animal's DNA (dissolving and digesting and removing all the other components that made up the animal). Next, using a magical technique called polymerase chain reaction (aka PCR, the basis of DNA fingerprinting from tiny quantities of DNA), Bourlat was able to pull out from the animal's entire genome a small number of gene sequences to compare with the equivalent genes from other animals.

The final product of a PCR experiment is a droplet of liquid inside a small plastic tube (just like the ones used for covid lateral flow tests). The tube contains trillions of copies of the tiny bit of DNA you want to analyse. But these copies are still swimming in a soup made up of all the other fragments of the animal's genome that went in at the beginning of the PCR experiment. To separate out the PCR-amplified fragment of interest, all the DNA from the tube is pipetted into a small depression in a slab of agarose gel (very like jelly), and an electric field is applied – essentially the same process of electrophoresis that Carl Woese used to separate bits of DNA/RNA of different sizes back in the 1970s. At the end of this process there will be fragments of every size smeared across the gel, but the PCR-amplified fragment of a particular size will be present in vast quantities and therefore easily visible on the electrophoresis gel.

When we did the experiment using the DNA extracted from an entire animal (gut and all), we found high concentrations of DNA fragments of two slightly different sizes – our PCR experiment had amplified two versions of the gene we were interested in. Finding two different versions of a gene is exactly what you might expect if you had amplified the gene from a sample that contained DNA from two different species. When we repeated the experiment using the DNA from the sample with the animal's gut removed, one of the two bits of DNA – let's call it gene version 1 – had disappeared. Because gene version 1 disappeared when the animal's stomach was removed, we had effectively shown that it came not from the animal itself but from a creature it had eaten. The next part of the experiment was to find what kind of animal the DNA found in the gut came from. As we had suspected, the DNA proved to be identical to the mollusc-like DNA sequence that had been published. We had shown that the published result was a mistake, the DNA a contaminant; *Xenoturbella* is not a mollusc, but it does eat them.

So far so good, and everyone *still* agrees with us that *Xenoturbella* is not a mollusc. But we still had the remaining gene version 2 to look at. This is the gene that was found in both experiments, gut or no gut. Gene version 2 really must be one of *Xenoturbella*'s genes, so comparing it to the equivalent gene from other animals should finally tell us where in the animal kingdom *Xenoturbella* really belongs. Our results showed, to everyone's surprise, that *Xenoturbella* is related to the deuterostome animals – the vertebrates and starfish; this unimpressive little worm is (I still believe) one of the closest relatives of the vertebrates and an important source of clues to our early evolution.

Again, twenty years on, I find I cannot reconstruct the series of events surrounding this discovery; what I had expected to find at the time; what it was like as we did the analysis – a sudden, exciting *a-ha!* or a gradual unfolding of an amazing new result. The one thing I do remember is worrying a lot about getting it wrong and that Bourlat carefully checked and rechecked that the mollusc DNA really was the contaminant. The paper reporting the work was, like the original papers claiming a mollusc relationship, published in *Nature*, and we gave it the plain-English title '*Xenoturbella* is a deuterostome that eats molluscs'.[5] (*Nature*'s rules against certain forms of punctuation in titles prevented us from adding ': the diet of worms'.) For a few happy years this was a generally accepted result.

This all changed (cue the sound of a needle scratching across a record) when it became obvious that *Xenoturbella* was not alone on its twig on the tree. We have discovered, using more DNA sequences, that *Xenoturbella* is in fact closely related to the acoelomorphs (the simple worms that include the mint sauce worm).

Uniting these two groups of simple worms has been the cause of two decades of argument. The argument exists because the acoelomorphs were thought to belong somewhere else on the tree entirely – right at the bottom of the bilaterian branch.

While *Xenoturbella* and the acoelomorphs clearly are close relatives (their branches have been combined into a phylum called Xenacoelomorpha), the two different places they had appeared to occupy on the tree of life set up a struggle of the 'xenacoelomorphs are a simple branch at the bottom of the animal tree' school (bad) versus the 'xenacoelomorphs are related to the starfish and vertebrates' school (good, obviously).

I think it is worth stepping back here to ask why – beyond wanting to know the structure of the tree and perhaps giving these branches a suitable name – we might actually care where Xenacoelomorpha belongs. The significance (I hesitate to use the word 'importance') of resolving this problem is inherent in all that we have seen before; knowing the true shape of the tree is the key that allows us to reconstruct the past. Let's look, then, at the evidence for two rival trees – I will try to give a fair account for the version that I disagree with – and ask what their different implications for reconstructing our ancestors are.

The first tree puts these worms close to the bottom of the trunk of the animal tree; they branch away as an independent line of evolution before the first appearance of all the other bilaterally symmetrical animals. This position certainly makes sense of the set of characters that they lack (separate mouth and anus, a brain, organised gonads, coeloms, kidneys, etc.). This tree gives a parsimonious explanation of the evolution of these characters saying that they each evolved just once in the branch that leads to all bilaterian animals except the xenacoelomorphs, which left the party too early.

The morphological characters are not the only things that seem to be missing in the xenacoelomorphs. In most bilaterians, there are roughly eight Hox genes – these, remember, are the famous genes that tell the cells in an animal's body where they are on the head-to-tail axis (and which, when mutated in a fruit fly, produce four wings instead of two or legs where antennae should be). In the xenacoelomorphs, however, there are only a maximum of five Hox genes. This small number has been interpreted as representing an intermediate stage in the evolution of the Hox genes before the full set of eight appeared. This all makes good sense – a simple body and a simple set of the genes that make that body are easily explained by xenacoelomorphs being an early branch of the animal tree of life.

At this point, it is worth circling back to the argument about the age of the bilaterian animals. Remember that molecular clocks seem to tell us that the ancestor of the bilaterians is much older than the first bilaterian fossil. And one of the explanations for the millions of fossil-free years was that the bilaterian ancestors were tiny, simple and insubstantial.

The xenacoelomorphs seem almost too good to be true for this view of the Precambrian animals. They are generally tiny, very simple and soft, with none of the easily fossilised hard parts like shells or jaws or a carapace that are found in other bilaterian groups. They are a convenient model for the first of the bilaterians, and their biology could conveniently explain a 120-million-year absence of fossils.

It gets even better, because ancient xenacoelomorphs would also demonstrate the evolutionary steps that led to the first bilaterians. Without the xenacoelomorphs, we have an evolutionary chasm between the complex bilaterians and their closest relatives – the very simple jellyfish and sea anemones. The xenacoelomorphs would step into this gap: more complex than a jellyfish but not as sophisticated as a fish or mollusc or insect; more Hox genes than a jellyfish but not the full set; bilaterally symmetrical but still missing certain advanced features such as a brain, gonads, coeloms and so on. If my frenemies are correct about xenacoelomorphs belonging at the bottom of the tree, they would make a wonderful missing link.

The alternative tree has implications for animal evolution that go in completely the opposite direction. Instead of interpreting the xenacoelomorphs as having changed very little in over half a billion years, the second tree actually requires that they have changed *much more* than most animals. This tree – which sticks the simple xenacoelomorphs right inside the complex deuterostomes near to starfish and humans – is the one that I believe to be correct.

If I am right and the xenacoelomorphs do belong plonked in the middle of the deuterostomes, then the story of their evolution must be one of loss. Having evolved from the same complex ancestor as all the other complex animals, they must at some point in their past have possessed all those complexities. To explain why they look so simple now, we have to conclude that they *lost* all of these characters for reasons unknown. My proposed place on the tree of life for these worms demands this less parsimonious explanation, in which all of these characters had to evolve once (the same as on the other tree) but then all had to be lost in order to end up with the simple xenacoelomorphs alive today.

Fortunately for me, this isn't an entirely hopeless case, not least because we already know that evolution can go in reverse, that organisms can lose characters – look behind you to see if you still have a tail. The flatworms themselves, the lookalikes with which the xenacoelomorphs were for a long time linked, have shown us an excellent precedent for this explanation. The idea that xenacoelomorphs might also have evolved from complex large animals is clearly not the simplest explanation of how they look like they do today, but it is far from impossible.

The very first paper we published on *Xenoturbella* was widely covered in newspapers and on radio, and Bourlat was even interviewed on the BBC TV lunchtime news. The *Sun*, the infamous British tabloid newspaper, covered our story from its own unique angle: 'Humans are related to a weird worm with no brain . . . And the link may explain the slimy wriggling during the Iraq War of French President Jacques Chirac, famously dubbed Le Worm by The Sun.' We even made it into the *Sunday Telegraph* magazine 'A–Z of the Year 2003', maybe helped a little by the lack of notable events beginning with an X (Rugby World Cup-winning Johnny Wilkinson was 'W'). The media interest was due to our discovery that a drab little worm is one of humanity's closest invertebrate relatives.

Xenoturbella (and its acoelomorph relatives) really are in my opinion most closely related to the starfish, sea urchins (and the hemichordate worms which I have steadfastly ignored). And therefore, following Grobben, closely linked to the vertebrates. This relationship tells us that comparing vertebrates to these animals is the best way to discover what the common ancestor of all deuterostomes was like. Xenacoelomorphs, starfish/sea urchins and hemichordates are each pretty weird on their own but, with luck, each one of this band of oddities will give up its own slightly garbled version of the story of the deuterostome ancestor; piecing these stories together should allow us to travel back in time to get a glimpse of the very first vertebrates.

In the next and final section, we are going to embrace this natural interest in our own distant origins; starting at the beginning with LUCA, we are going to climb the tree through the billions of years of our own evolutionary story. At each fork in the tree we will choose the branch that leads towards us. Our journey will take us through a series of our most important ancestors and will let us follow the steady accumulation (and occasional losses) of the many characters that were assembled by evolution to make us human.

PART III

Tracing Our Family Tree

18

The First Three Billion Years

THIS BOOK WAS never intended to be a description of the whole tree of life, but it would still be a little odd to reach the end without revealing what the tree looks like. Our natural limitations leave us with a problem, however, because it is far too big, and far too complex, for a human mind to grasp any more than its vague outlines.

To get an inkling of the scale of this problem we might compare the tree of life to a magnificent full-grown oak tree, perhaps one that sprouted a thousand years ago from an acorn buried by a squirrel at the time of the Norman Conquest. Imagine that this grand, forty-metre-tall oak tree is a tree of life. We could climb this tree, shimmying up its thick trunk and following a multitude of different routes upwards. At every branching point we would pass an ancestor – the source, at one point in time, of a pair of new lineages – and we can choose to follow either of the two smaller branches it gives rise to. As we climb, the branches get thinner and more numerous – we move from huge kingdoms to phyla, to classes and families, ending finally with single species represented by one of the million different leaves that a full-grown oak can carry. But here we encounter a large (and possibly hard to grasp) discrepancy in size and complexity between this realistic, million-leaved tree and the tree of life. A very cautious estimate of the number of species alive today lands around the 9 million mark.[1] An oak large enough to carry these 9 million leaves would be a freakish giant; at roughly 120 metres tall it would edge higher even than the tallest giant redwood we have ever measured. But this is just a starting point. To encompass the most extreme recent estimate for the number of living species – coming in at a round trillion[2] – we must imagine standing in the shade of an oak whose highest leaves tower 40 kilometres above us, halfway to space. The fact is that the dimensions of the complete tree of life entirely swamp our simile, vastly outgrowing the undoubted talents of the form of a tree to convey the events of evolution.

Our desire to stand back and survey the shape and structure of the tree of life can be rescued by a simple truth which is that almost every one of its embarrassment of details is of little interest to non-specialists. Most people – and I happily admit I am one of them – will not care about the relationships between the many millions of living species of Eubacteria and Archaea, or even about the 350,000 living species of beetle.

This convenient truth gives us another way to use this immense tree to travel through time. Our simple solution is simply to ignore all but a single one of the 4 billion possible paths up the tree that lead to a species. I have, needless to say, chosen to follow the history of our own species as it climbed the tree of life. Our journey begins with LUCA; at every fork we come to, the branch we pick will be the one that leads towards the human leaf. This path will introduce us to a series of ancestors each more closely related to us than the last; each with a little more in common with us, with a few more of the characters that add up to make a human.

While I've proposed a solution to the unimaginable complexity of the branching pattern of the tree of life, there is another dimension where we need a guiding hand – and this is time. Our human minds, limited by our lived experience of four score years and ten, find it hard to grasp the meaning of the millions and billions that encompass evolution. The problem of timescale is like trying to grasp the size and shape of continents from the point of view of your local park: to picture the whole world you need a globe. Rather than a map, I am going to use the familiar proportions of a human body to measure our journey up the tree of life. On this figure, the origin of the earth 4.5 billion years ago is positioned at the soles of the feet, and you and I (and everything else alive today) are perched on top of the head.

Before we go any further, I have to warn you that this human-centric approach risks making evolution look purposeful. Following our own evolutionary path could be read as a series of choices whose result is the steady accumulation of the parts and features needed to produce a human being – like the deliberate buying of ingredients for a recipe. The humbling reality is that evolution has no aim; the evolution of humans had no inevitability at all, and our path up the tree could have been snuffed out on a billion different occasions. A further danger is that our focus on ourselves relegates the rest of life to items of food or annoying parasites, to companions or just scenery in a human story. This would be pure vanity. On the vast scale of the tree of life, our species is a single leaf, a very recent arrival, and one that may not be around for long. When our leaf does fall, the tree of life will go on growing as little affected by our departure as it was by our arrival.

Warnings delivered, let's prepare to start climbing.

We started with classification, and, coming full circle, classification provides the natural structure to understand our place on the tree of life. Like all other species, humans belong to a series of branches that encompass ever smaller and more exclusive groups, starting with the trunk of the tree (encompassing all of life), finding a path to mammals, primates, apes and then, finally, the single leaf that defines our own genus and species. You may be surprised (and relieved) to discover that, despite the vastness of the tree of life, the list of ancestors that separate *Homo sapiens* from LUCA is incredibly short.

In order to explain just why our list of ancestors is so short, I am going to begin by making one bold assumption: I am going to pretend that the tree of life is perfectly balanced, so that at every branching point throughout the tree there is the same number of species on each of the two resulting branches. With this little fudge to make the maths easy, we are in a position to add up the number of individual ancestors a species has. The calculation can be understood by starting with a few species and building up from there: on a tree with two species, A and B, A has a single ancestor that it shares with B (the arrow in the diagram); for four species, A to D, A has two ancestors, sharing one ancestor with its closest relative B and a second older ancestor with B, C and D; for eight species, A to H, A has three ancestors; for sixteen species, A has four; and so on. (While I have focussed on species A, each species has the same number of ancestors.) What hopefully jumps out is that the number of ancestors grows very slowly as we increase the number of species. While ancestors get *added* – 1, 2, 3, 4 – the number of species gets *multiplied* – 2, 4, 8, 16.

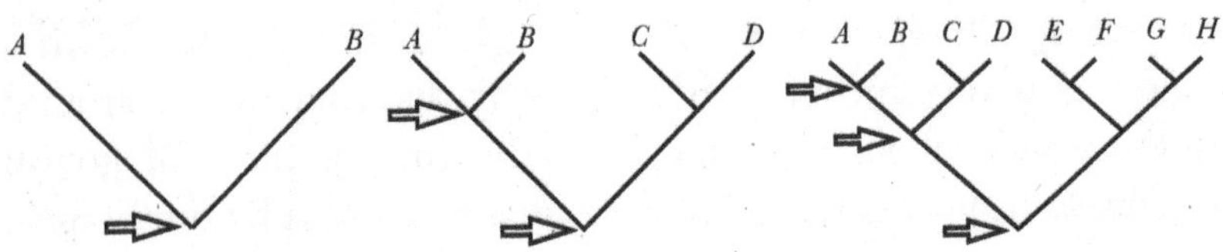

FIGURE 9: On a tree of two species, species A has a single ancestor (arrow). With four species there are two ancestors (arrows). With eight species, three ancestors.

From this we can make a bit of a leap and calculate that on a tree that links the 4 billion species that (at a conservative estimate) have ever lived, each leaf would be separated from the universal ancestor by just thirty-one ancestors. Each species, in other words, belongs to just thirty-one increasingly exclusive, nested groups of relatives, starting with the all-inclusive group that emerges from the root of the tree.

The USA-based National Center for Biotechnology Information (NCBI) provides a semi-official classification of life (or at least of the life that is represented in its databases of DNA and protein sequences), and this is as good a guide as any to the series of groups that humans belong to. The list is (as we have seen) shorter than one might have expected, and although even this short list is probably too long for the purposes of being interesting per se, I still want to give it in full. There can be a value in a boring list. The tedium of the 'catalogue of ships' in the *Iliad* serves to convey the scale of the Trojan War – here the list of our ancestors and the branches to which we belong shows that for the handful of hopefully familiar levels of classification, there are many more known only to the initiated.

Our list is as follows: Cellular Organisms (all the descendants of LUCA); Eukaryota; Opisthokonta; Metazoa; Eumetazoa; Bilateria; Deuterostomia; Chordata; Craniata; Vertebrata; Gnathostomata; Teleostomi; Euteleostomi; Sarcopterygii; Dipnotetrapodomorpha; Tetrapoda; Amniota; Mammalia; Theria; Eutheria; Boreoeutheria; Euarchontoglires; Primates; Haplorrhini; Simiiformes; Catarrhini; Hominoidea; Hominidae; Homininae; Hominini; *Homo sapiens*. A few groups are familiar, I am sure (Vertebrata, Mammalia, Primates), but I suspect they are a small minority.

We are going to start at the beginning of this list, with LUCA, the 4-billion-year-old root of the tree. From here we will climb on up until we reach the human leaf at the tree's top. We will pause on our journey to meet fourteen of our ancestors, each representing a crucial moment of our evolution.

LUCA, 4 Billion Years Ago

LUCA can be found low down on the calf of our human body timescale – roughly 4 billion years in the past. This moment in time marks the division between the end of the first aeon of the planet, the Hadean, and the onset of the second, the Archean (marked by the very oldest rocks we can find today). The earth, over the half a billion years since its formation, had cooled sufficiently for a solid crust to form over its molten interior and for the water to condense to form oceans. The atmosphere, pumped full of volcanic gases and containing 100,000 times less oxygen than it does today, would kill a human within seconds. The young sun was feeble, 25% dimmer than the one we enjoy, but a strong greenhouse effect kept the planet warm and the oceans liquid. LUCA's contribution to our own biology is immense, bequeathing to us (and to all of life) the most fundamental parts of our biology: DNA, RNA, proteins and much of the cellular machinery needed to synthesise, manipulate, copy and repair all of these; reproduction; the genetic code; many aspects of our cellular metabolism; ATP as the currency of energy; and, most extraordinary of all, the very fact of being alive. While it marks the first step of our journey, we won't dwell on LUCA, the sketchy details of whose biology we have thought about before. But, by its very definition, LUCA lies at the point on the tree of life where it splits to create two great branches, Eubacteria and Archaea, which will come together again in the form of our second step on the tree – the first eukaryote.

The First Eukaryote Cell, 2.2 Billion Years Ago

Arriving here requires a huge 1.8-billion-year leap in time to land in the early part of earth's third aeon – the Proterozoic. We are only at the second stop of our journey, but we have already reached the hips of our human-shaped timescale. Across the huge span of time that this represents, the archaeal and the eubacterial branches of life diversified into many individual branches, spreading across the globe and adapting in diverse directions to occupy every exploitable niche.

The geology of the earth changed as well; much more land had risen above the surface of the oceans, and the newly grown continents were surrounded by shallow seas. But perhaps the event with the largest impact on the planet and on life was the evolution of photosynthesis in one group of bacteria – the cyanobacteria. Over hundreds of millions of years, these tiny cells filled the atmosphere with oxygen in a slowly unfolding catastrophe known, prosaically, as the Great Oxygenation Event.[3] It is thought to have killed off a significant proportion of life on earth,[4] but the poisonous gas was also the urgent spur driving the union between an oxygen-hating archaeal host cell and its oxygen-loving bacterial guest. The eukaryotes that were the result of this merger gave rise to all complex life alive today: from tiny amoebae, ciliates and countless other single-celled protozoan species to the large and many-celled fungi, plants and all the animals including ourselves.

While there may be a passing interest in the diversity of life that exists within the two ancient branches coming from LUCA, I want to focus on finding out which of the many lineages *within* the archaeal and bacterial branches were the ones that merged to make the first eukaryotic cell. By finding the closest living relatives of these two ancestors we hope to discover something about the lives and characteristics of the two founders of our family. The biology of these living relatives must contain clues explaining what it was that first attracted one to the other and why their marriage turned out to be so very successful.

It is very hard to study a single bacterium. The problem is their size – you could fit a billion of them in a space the size of a sand grain. Their tininess makes it all but impossible to analyse what they are made of, like trying to taste a cake from a single crumb. This simple, stubborn limitation meant that, for hundreds of years following their first discovery, our knowledge of Eubacteria and Archaea was restricted to the few species that could be cultured in large numbers in the laboratory. This method for studying bacterial biology and diversity was obviously severely limited, just as studying the birds that visit my garden would give me a feeble grasp on the diversity of birds around the world. As recently as forty years ago, the number of known bacterial species was in the low thousands, but this situation was transformed over the course of a few short years when new technology made it possible to find species not only by observing them directly but by detecting their DNA molecules within their natural environments.

In the late 1980s, building on the work of Carl Woese, microbiologists began to look beyond the few bacteria that were happy in a petri dish or test tube. They sampled oceans, lakes and rivers, the spoil left over from mining, and the water trickling though the soil of a rainforest. They sampled water coming from tens of metres below the earth's surface (famously in the cold sulfidic groundwater emanating from a man-made spring – the wonderfully named Mühlbacher Schwefelquelle) and literally thousands of other environments.[5] From each of these samples they would separate out the tiny quantities of DNA from whatever community of organisms might have been hiding there and use the newly invented PCR method to transform a tiny handful of interesting DNA molecules into trillions of easily analysed copies. This mix of DNA molecules from many species could now be searched for each individual species' own version of the universal gene (SSU rRNA) that Woese had studied.

What they found in these samples was tens of thousands of different SSU rRNA genes. Each one of these was clearly related to the known eubacterial (or archaeal or eukaryotic) genes, but, crucially, almost all of them were different from anything they already knew about. These new genes were evidence of species that had never been seen because they couldn't be cultured. Four decades later, the number of bacteria we know of has grown roughly a thousandfold to more than one million distinct species. The number of known archaeans (which are even harder to grow in a petri dish due to their extreme lifestyle preferences – very hot, acidic, oxygen-free, sulphur-rich, salty, alkaline or deep) has blown up even faster.

Our search for the two species that merged to make the first eukaryote takes us, first of all, in amongst the branches of Eubacteria. We are looking for the species that became engulfed and which persists today in the form of the eukaryote mitochondria. The relationships between the million known eubacterial species remain a topic of robust debate, but there is broad agreement over the big picture. There are two huge branches of Eubacteria: Terrabacteria (named because many are terrestrial) and Gracilicutes (from 'gracile', meaning 'slender').[6] Within these two great branches, roughly 100 smaller branches can be distinguished, and these may be more different from one another than the animal kingdom is from the plant. More different, that is, than a frog is from its lily pad.

Our real interest is not (thank goodness) in meeting these hundred branches, but in looking amongst them for the identity of the bacterial half of the eukaryote – the invader that became our mitochondrion. In a perfect world we would be able to narrow this down not just to the closest bacterial kingdom (amongst a hundred others) but eventually to the single closest living branch.

The microbiologists seem, to an outsider, to be a querulous lot. There is some dispute over the closest living bacterial relative of our mitochondria, but it is generally agreed to lie somewhere amongst (or at the very least immediately adjacent to) a group called Proteobacteria that lies within the gracilicute half of the bacterial tree of life. The proteobacterial branch was first recognised as a separate entity long ago by Carl Woese.[7] Its name, coming from '*Proteus*, a Greek god of the sea, capable of assuming many different shapes',[8] recognises the great diversity of forms taken by the Proteobacteria,* and this can probably best be interpreted as telling us that there is not much, beyond their similar genes, that obviously unites them. Proteobacteria contains a good dose of species that, if you have been unlucky, may be intimately familiar to you. Gammaproteobacteria includes *Escherichia coli* – usually harmless and present in its trillions in your gut – but also *Salmonella*, *Yersinia* (cause of the plague) and *Shigella* (which causes perhaps 100 million cases of diarrhoea a year and kills hundreds of thousands, mostly children). Mitochondria turn out to be most closely related to Alphaproteobacteria, which includes *Brucella* (the cause of diseases with intriguing folk names such as Brucellosis, Bang's disease, Rock fever of Gibraltar, slinking of calves and ram epididymitis) and *Rickettsia* (cause of African tick-bite fever, Rocky Mountain spotted fever and Flinders Island spotted fever). *Rickettsia*

* To the dismay of some traditionalists, a revamp of bacterial branch names in 2021 changed the name from Proteobacteria to Pseudomonadota.

may be the most intriguing of all of these because, like our mitochondria, it spends its life as a guest inside other cells.

While the argument about exactly which of the different groups in Alphaproteobacteria is closest of all to the eukaryote mitochondria continues, it is in the biology and genetics of these nearest relatives of the mitochondria that we are now looking to understand what attributes it had that made the endosymbiosis possible. The second half of the eukaryote puzzle is finding the closest living relatives of the 2.2-billion-year-old archaeal cells that swallowed this proteobacterial passenger, and this search takes us to a tiny volcano at the bottom of the sea.

Once, as a team-building exercise, I took the members of my lab across the English Channel in a sailing boat; the sea was pretty rough, the wind was strong, and my stomach was weak. My intention to lead my team having flown out the porthole, I spent several humbling hours lying on the bottom of the cabin in a puddle of water. My lamentable sea legs put me in awe of the marine biologists and oceanographers who spend days or weeks on research boats in much rougher waters. In 2008, a team of scientists aboard such a boat – the Norwegian research vessel *G.O. Sars* – made an amazing discovery in a part of the ocean that fills me with horror.[9] They were geologists, not biologists, and their finding was of undoubted interest to their fellows; but what they found would soon lead to a discovery that changed our ideas of the origin of the eukaryotic host cell.

The expedition, led by Professor Rolf Pedersen, went looking for undersea volcanic activity in the vastness of the Norwegian Sea – in the approximate middle of a triangle formed by northern Norway, Iceland and the island of Svalbard.[10] Describing what they discovered as 'volcanic activity', if you are picturing an underwater Mount Fuji, is terribly misleading. The volcano they found was a small constellation of hydrothermal vents – tubes like organ pipes made of minerals that had precipitated out of the volcanically super-heated water gushing out of a fissure in the sea floor. The very biggest of these was a narrow pipe 13 metres tall, a minuscule object to locate 2.5 kilometres below the surface of an endless sea. Tracking down these pipes required equipment able to detect the subtle whiff of the hot, mineral-rich water that pours out of them (to be instantly diluted by quadrillions of gallons of cold seawater). Their equipment was sensitive enough to detect a plume of water that was warmer than the rest of the ocean by just one ten-thousandth of a degree. It was a search that was at times so confusing, thanks to capricious currents disturbing the plume, that they named the site they discovered Loki's Castle to honour the shape-shifting Norse god Loki. The biological thunderbolt we are interested in came a few years later when the life forms in the sediment around the pipes of Loki's Castle came to be studied. What the biologists found was an entirely new group within Archaea that they named Lokiarchaeota.[11]

When Carl Woese first discovered the enormous gulf between the domains of Eubacteria and Archaea, he recognised, within the latter branch, two new kingdoms: Euryarchaeota, the Greek *eury* meaning 'broad', reflecting the many niches the species occupied, and Crenarchaeota, *cren* meaning 'spring' or 'fount'.[12] The hot-water environments the Crenarchaeota live in seemed plausible as the conditions favoured by the oldest archaeans. Thanks to years of work and many environmental DNA studies over subsequent years, we now recognise four main branches of Archaea: Woese's Euryarchaeota, two new groups called Thermococcales and DPANN (one of the horrible, unmemorable acronyms named for the smaller branches within it that microbiologists seem to love), and finally another large, acronymed group called TACK which includes Woese's Crenarchaeota – the 'C' of the name.[13]

Comparisons of SSU rRNA genes had long suggested that the TACK branch of Archaea is where we might find living relatives of the cell that made the first eukaryote, but the link had always been a bit vague; the discovery of Lokiarchaeota changed this overnight. The new species was not immediately seen in the flesh, but its existence was inferred from the DNA found in a sample of sea-floor sediment from the periphery of Loki's Castle. These

DNA molecules were clearly from a group of TACK archaeans, but they were different from any that had been seen before. A new branch of archaeans is of undoubted parochial interest, but the climactic revelation (and the thing we are seeking in our journey towards humans) was that Lokiarchaeota is clearly closer to the eukaryotes than any other archaeans.

Lokiarchaeota has gone on to become a priceless resource for discovering what the very first eukaryotes looked like. Its first discovery was followed by a rush of related archaeal species, some of them even more closely related to the eukaryotes than those from Loki's Castle. These additions have also been given Norse names – Thorarchaeota, Odinarchaeota and Heimdallarchaeota – and the superphylum containing all of these has been named Asgard after the realm of the Norse gods.[14]

Microbiologists are now in a position to ask questions about modern asgardians: how they get their energy to survive; where and in what conditions they live; what important genes they share with eukaryotes (it turns out to be more than any other archaean); and whether these genes might have been essential for the evolution of the unique features of living eukaryotes that make them so much larger, more versatile and more complex. It has, in short, become possible to start asking what it was about both the archaean asgards and their partner within the alphaproteobacteria that allowed them to join forces so successfully – an event so utterly improbable that the earth had waited at least 2 billion years to witness it.

The Opisthokont Ancestor, 1.3 Billion Years Ago

Since their origin in a marriage between an archaean and a bacterium, the eukaryotes have flourished and diversified spectacularly, going on to become some of the most important organisms in almost all ecosystems. Their diversification and spread took place throughout the Proterozoic era amid huge tectonic shifts, with continents coalescing and splitting apart and moving from tropics to poles and back again. Oxygen levels in the atmosphere increased, and the planet spent long periods almost totally frozen over, like a giant snowball. While eukaryotic diversity today is most immediately recognised in the large forms of many-celled organisms like animals, plants and some fungi, eukaryote roots go much deeper than these few show-offs; the real diversity is to be found in several supergroups (really big branches) of single-celled creatures commonly referred to as protozoans.[15]

The TSAR supergroup includes dinoflagellates, known for making 'red tides' – toxic algal blooms that can kill sea life and cause food poisoning in people eating seafood from the area. TSAR also includes foraminiferans, beloved of oil prospectors because their distinctive and abundant shells disclose the exact ages of the rocks they are fossilised in. The Haptista supergroup includes coccolithophores, whose tiny, calcium-rich shells drifted down to the bottom of a Cretaceous sea in such astounding numbers and for so long that they produced the gigatonnes of chalk in the white cliffs of Dover. The Archaeplastida supergroup of algae and plants was founded when a second endosymbiosis occurred about a billion years ago; this involved one group of eukaryotes engulfing a second bacterium – a photosynthesising cyanobacterium, which, as the dreadful Konstantin Mereschkowski had proposed, became a chloroplast. This second merger established the branch of the tree that is the ultimate source of all animal food. Animals themselves, meanwhile are just one part of a group called Opisthokonta, itself a branch within a supergroup called Obazoa. The opisthokonts first appeared about 1.3 billion years ago, and we have reached the solar plexus of our human timescale.[16]

This name 'opisthokont' comes from the Greek for 'rear-facing spear'.[17]* The 'spear' refers to the flagellum that sticks out of the back of opisthokont cells to propel them forward – if you think

* The name was first used by the Swiss biologist Wilhelm Vischer in 1945 for one of the species that make up the group, and new branches, including the animals, have been added (and occasionally taken away) over subsequent decades.

of the tail at the back of a human sperm cell you will get the picture. The only opisthokont group that is likely to be familiar (besides the animals) is the fungi, a group which, for most of human history, had been lumped in with plants and studied by botanists. But fungi also have sperm-like cells with the same rear-facing flagellum as animals.

Though the fungi are the most familiar of the opisthokonts, our closest relatives in the group are called Choanoflagellatea. These (usually) single-celled organisms have a typical oval body with a small collar at one end made of tiny struts which support a membrane. The *choano* in 'choanoflagellate' means 'funnel' and refers to this collar structure. In the middle of the collar there is a long flagellum, pretty much indistinguishable from the tail of an animal sperm. Surprisingly, the close relationship between the tiny choanoflagellates and the animals was first spotted as long ago as the nineteenth century – a discovery usually credited to the British biologist William Saville-Kent.[18] Saville-Kent found that sponges (a simple-bodied animal) have a type of cell (called a choanocyte) that, just like a choanoflagellate, has a flagellum surrounded by a collar. These cells, of which sponges have a huge number, all cooperate to waggle their flagella in the water to produce a strong current, and it is this current that sucks food particles into the sponge where they are captured and digested.

Our climb up the tree of life has led us already to the cusp of the animal kingdom. We have arrived here in two giant leaps. Our starting point, low down the leg of our human timescale, was LUCA – the universal root of the tree of life, the origin of everything alive today and the very closest we can get to the origin of life. Our first jump took us forward 2 billion years, to the hip of our human timescale, to meet the first of the eukaryotes. This individual, formed in an extraordinary fusion of two species of tiny, simple prokaryotes, produced an organism that was much more than the sum of its parts. This new form of life – the founder of a new domain on the tree of life – passed on to all of its descendants a complex and relatively huge oxygen-breathing cell with a nucleus and chromosomes and sex; flagella and cilia to swim; mitochondria to provide the energy needed to run this complex machine; the ability to make pouches in the cell's membrane to eat and to excrete; and many more arcane characteristics besides. Our second step (a move up to the solar plexus of our human timescale) covered another 1.2 billion years and took us to our own local group of single-celled eukaryotes, the opisthokonts. The very oldest animals are close by and, in the next chapter, the legacy of the opisthokont branch will become clear. We are going to climb on up the tree, the ancestors now coming thick and fast. Our journey will take us from the first and simplest of the animals to the very last (and arguably most complex) in the form of our own leaf on the tree.

19

The First Animals

IN 1665 THE recently founded Royal Society of London (for Improving of Natural Knowledge) published as its inaugural book the extraordinary *Micrographia* by Robert Hooke. *Micrographia* contains Hooke's exquisite drawings and detailed written observations of diverse substances (silk, Muscovy glass, sand); of objects (the point of a needle); of parts of plants (poppy seeds, nettle stings); and of animals (insects, sponges, the teeth of a snail, feathers); all of them observed in extraordinary and unprecedented detail using his magnifying glasses and microscopes. Hooke's most famous image, of a flea magnified to 300 times its natural size, revealed to human eyes for the first time the astounding intricacy that characterises the body of even something so tiny. Hooke was able to describe the details of the flea's body, 'adorn'd with a curiously polish'd suit of sable Armour, neatly jointed, and beset with multitudes of sharp pinns, shap'd almost like Porcupine's Quills, or bright conical Steel-bodkins'. He found the minuscule perfection and complexity he could see in this divinely created animal to be incomparably finer than that quintessence of man-made precision – a needle – whose point his microscope revealed to have 'a broad, blunt, and very irregular end'.[1]

What Hooke's new instruments had revealed was that, close up, even the tiniest of the millions of animal species is an amazingly complex beast. And this has turned out to be true of every animal, from the 'are they even animals?', sea sponges and sea squirts, to octopuses and dragonflies; from fleas and vinegar worms to Nobel Prize-winning, *Magic Flute*-composing humans. The change from a one-celled ancestor to many-celled animals – a single step in our journey from LUCA to humans – was going to have the most extraordinary, unanticipated consequences, like the impacts on our history that rippled out from the invention of writing or the wheel. But how did this change come about? And why, of all the lineages of tiny, simple, single-celled eukaryotes, was it our opisthokont ancestors that made this leap in complexity? Did they have something special that made becoming an animal possible? We can't, of course, interrogate these ancestors directly – they breathed their last perhaps 700 million years ago – but looking at our closest neighbours on the tree of life, the choanoflagellates, we should be able to pick up some clues.

Before we put our simple relatives under the microscope, it is worth thinking about what it is that we are even trying to explain. What is it about animals that is so special? The biggest novelty is, simply, multicellularity – your own body contains tens of trillions of cells, for example; a blue whale may have 10 quadrillion. Around this single, central fact, everything else that appears most remarkable about animal biology radiates. Some of what the clique of cells that made the earliest animals had to do is rather obvious. First and foremost, they had to invent (by which I mean evolve) a way to remain stuck together after they had divided: a new gene was needed to glue the cells together, or perhaps an existing gene was repurposed (from a gene that we can search for in our single-celled relatives). To function as an organism rather than a collection of individuals, animal cells had to evolve ways to talk to each other, to coordinate their actions and to influence each other in useful ways. Related to this, and perhaps most interesting of all, an animal's many cells had to take on different and specific roles and then to behave altruistically, to cooperate – you go shopping, I'll cook and he can do the washing up. Animal cells have become super diverse: we have muscle cells and nerve cells, cells that line our stomachs and secrete digestive enzymes, cells that detect light or sound or temperature or movement or pressure or taste, cells covered in cilia that waft particles out of our lungs, blood cells that carry oxygen, lymphocytes that destroy bacteria and a thousand others besides. These many kinds of specialists cooperate, each contributing to the common goal by expertly performing their own allotted task, like workers on a production line. As the ultimate example of their altruism, and unlike any single-celled organism, almost every cell in the body of an animal has given up any possibility of passing its genes on to the next generation. This extraordinary privilege is reserved for the specialised germ cells: eggs and sperm.

With the evolution of multiple different kinds of cell there came a need to order them into larger structures: tissues (muscle, bone, blood, nerves) and organs (brain, kidney, stomach). Organising cells allows them to work efficiently; a body that jumbled up muscle, nerve and digestive cells would be a disastrous monster. Your muscles are concentrations of billions of individual muscle cells, and muscles work only because the individual cells have been carefully organised to lie side by side so that they all pull in the same direction. Your kidneys are made of millions of tiny filtering structures (called glomeruli), each contributing a droplet of urine to the overall outflow of waste; and each tiny glomerulus is itself composed of several different cell types each organised to sit in the right place according to its special role.

Finally, animal cells are organised to make a body with a certain shape. Our tissues and organs themselves are arranged just so in order to function together. Different bits of your body have to be scaled to the right size: left leg same length as right leg; heart not too big and not too small; a brain that is of a shape and size to fill your cranium; your circulatory system arranged to supply every one of your cells with food and oxygen. These higher levels of organisation required the invention of new genes to regulate embryonic development. As Hooke's discovery of the intricacies of a tiny flea showed, all of this new complexity (and the new genes this implies) needed to be invented simply to make a tiny insect. Everything else in animal evolution could be thought of as more or less subtle variations on this wonderful new theme.

To understand the roots of these animal innovations, we can

look for parallels and precursors in our closest single-celled relatives. The first is the ability of some of our near neighbours to form small colonies of genetically identical individuals.[2] Just like an animal, there is an initial cell that divides multiple times and the new cells that result then remain stuck together. These cell clusters are produced by cell division (mitosis) rather than by the gathering up of a bunch of unrelated individuals; and this means the cells of the mini colony are genetically identical and so are primed to cooperate with each other.[3] From a gene's point of view, cooperating with other cells in such a colony would be helping an exact copy of yourself. Different species of these 'single-celled' relatives even grow into tiny colonies with different morphologies, and the shapes these adopt depend not on the shapes of individual cells but on how these cells are arranged.[4] Some species form chains of cells, some are linked in branched chains, others make a ball of cells on a stalk and others make tiny rosettes. Some of these mini colonies look for all the world like the earliest stages of a growing animal embryo.

A slightly rarer though more striking animal-like trait found in at least some species of choanoflagellates is the ability of their cells to change what they look like, taking on different shapes and different functions.[5] This is a pretty amazing trick – like animals, they can use one set of genes to build more than one kind of cell: it's like using the same ingredients to make a pancake or a Yorkshire pudding. Under certain environmental conditions, choanoflagellates can absorb their characteristic collar structure and transform themselves into a blob-like cell (no collar, no flagellum, no oval cell body), and, rather than swimming like a sperm, this second cell type moves by flowing like an amoeba.[6] This seems to suggest the beginnings of an animal-like flexibility in the forms their cells can adopt.

Another way to compare animals with choanoflagellates is to think about the genes they might have in common, and choanoflagellates have recently been shown to have a healthy number of the genes that had previously been thought of as quintessential components of a multicellular animal's toolkit.[7] There are genes animals use to glue cells together, genes used in making embryos, genes involved in communication between cells, and so on.[8] At this point we must resist the temptation to conclude that our single-celled ancestor had been preparing itself to become an animal. Evolution has no ability to plan; the genes that ended up being co-opted for making animals must have had important (and probably rather different) roles in our non-animal ancestor.

The Animal Ancestor, 600 Million Years Ago

In our journey up the tree of life, we have arrived at the fourth stage – the ancestor of all the living animals. While we are a little uncertain exactly when this beast lived, we have climbed to a point approximately 600 million years in the past, close to the end of the Proterozoic aeon, and roughly at the bottom of the chin of our human timescale. The very oldest animal fossils appear in sedimentary rocks that were formed during this time (immediately following a brutal, 100-million-year ice age). The first animals to show themselves on the warming planet are named the Ediacaran fauna after the Ediacara Hills in Australia where their fossils were first discovered. Ediacarans take various simple forms – some are disc-shaped (resembling a jellyfish) or leaf-shaped (with repeating patterns like the frond of a fern); some are known only from the fossilised traces of their simple burrows (which imply the existence of a burrow-maker that moved like a slug). We are currently in the dark as to which of these very oldest animal branches made it through to today – it seems likely that most were optimistic experiments that had their day in the sun but, superseded by new arrivals, became as outdated as a buggy whip maker on the arrival of the Model T Ford.[9] We are on safer ground with our next move into the thicket of animal branches, which clearly did survive into the most recent period of earth's history – the Phanerozoic aeon. 'Thicket' seems appropriate at this point, because, while we know which branches from this time have survived to today, we are wading into a mess of arguments about the order in which they branched off from the animal ancestor. This uncertainty over how they are related makes it tricky to reconstruct the series of steps that led from the first of the animals to the ancestor of our next branch – the bilaterians.

We should start with what we agree on, though, and this is simply that there are five individual branches to be sorted out, and that these five separated from each other in the few tens of millions of years following the origin of the animals. The first four branches are each a separate phylum of animals, and two of these, the sponges (Porifera) and the jellyfish/anemones/corals (Cnidaria), may be familiar, often seen in a rock pool or washed up on a beach. The other two, much less likely to be casually encountered, are the microscopic 'plate animals' (Placozoa) and the comb jellies (Ctenophora), also known as sea gooseberries. The last and by far the biggest of the five branches, itself containing about twenty-five individual animal phyla, is Bilateria.

Amongst these five, the sponges have long seemed the most plausible first branch, most distant from all of the others; this is certainly what Linnaeus believed, classifying the bath sponge, *Spongia*, within the plant kingdom, alongside the algae. Sponges certainly lack some fundamental characteristics found in most other animals – most notably they have neither muscle nor nerve cells – consistent with sponges separating from the main trunk of the animal tree before these characters evolved. While dogma for at least a hundred years, this idea has become one of the most enthusiastically fought-over parts of the animal tree in the past fifteen years, the argument making its way onto BBC radio ('a ferocious argument between two groups of scientists') and the pages of the *New York Times* ('A battle is raging in the tree of life').[10] The controversial alternative to sponges as the first branch of the animal tree – and this idea, as I write, is in the ascendancy – is the comb jelly branch. This odd idea seems to tell us that the super-simple sponges have thrown away all of the advanced characteristics – nerves, muscles, eyes, a symmetrical body – that the comb jellies, jellyfish and bilaterians have in common.

Loss of characters is, as we have discovered, a frequent event in evolution, and another great example can be found in the placozoan branch; these are surely one of the weirdest of all the animals. The first of just four known species of placozoans was discovered in 1883 (by chance in a seawater aquarium).[11] The scientific name of this species is *Trichoplax adhaerens*, which translates as 'hairy plate that sticks'. The 'hairiness' refers to the covering of tiny cilia (those miniature versions of a sperm flagellum) that stick out of the cells of *Trichoplax*, wafting around below its body and allowing it to move. The discoverer of *Trichoplax*, the German zoologist Franz Eilhard Schulze, described its simple body: a very small number of different kinds of cell; the top surface of the 'plate' distinct from the bottom but having no other body axes – no front or back, left or right; and no organs, no mouth, no nerve cells, and so on.[12] What *Trichoplax* is, where it fits into the story of animal evolution, was for a long while very unclear (and, to be honest, not exactly a hot topic). Schulze himself could find no obvious link to any other animal group, not to the sponges, comb jellies, jellyfish or bilaterians, but guessed, accurately as it turns out, that the placozoans are an isolated branch located somewhere near the root of the animal tree of life.

Recent work tells us that placozoans are related to the relatively complex jellyfish and to the very complex bilaterians, and there are other hints that they are (or perhaps I should say used to be) more complex than they look.[13] In a delightfully quirky set of experiments, it has been shown that the very short proteins that complex animals use in nervous signalling (a familiar human example would be the 'love hormone' oxytocin) also exist in the placozoans. Bathing placozoans in these 'neuropeptides' affects their behaviour in odd ways – causing them to twirl around, to crinkle up or to flatten themselves – they can, in other words, detect these neuropeptides.[14] Is this evidence of a primitive nervous system? Or are these behaviours what remain of a once complex nervous system that has lost its way?[15] Despite its simplicity, *Trichoplax* seems to be more closely related to the fairly complex jellyfish than to the sponges or comb jellies. The best bet is that the simple body of placozoans doesn't give us an insight into an early stage in animal evolution, but rather is another case of evolution in reverse – a simple animal descended from a complex ancestor.

The Bilaterian Ancestor, 555 Million Years Ago

There is (you may recall) some uncertainty about the date of the next stop on our journey towards humans: the ancestor of the bilaterians. Urbilateria must have existed before the beginning of the Cambrian period (538 million years ago), when a disparate band of bilaterian species burst onto the fossil record. To a good enough approximation, we are halfway up the chin of our human timescale's head. The trigger for the explosion of bilaterian life has been sought in many places, but the truth probably lies in some combination of late Ediacaran planetary conditions and new biology. The world was warm, the snowball earth long since melted; several continents were clustered together in the southern hemisphere, making a supercontinent called Gondwana; the amount of oxygen in the atmosphere and in the shallow regions of the seas had increased. These geological opportunities may have coincided with any of a number of biological innovations (eyes, shells, Hox genes, bilateral symmetry, separate mouth and anus and so on) to spur the rise of the bilaterians.

The bilaterians are easily recognised by their most obvious characteristic of having mirror-image left and right sides. This contrasts with all other animals outside Bilateria: sponges and placozoans have a top and a bottom but no consistent symmetry to their bodies; the comb jellies have a top and bottom but they can be sliced top to bottom along two axes of symmetry (think of a rectangular lemon drizzle cake that can be split into mirror halves either lengthways or widthways); jellyfish and sea anemones, finally, have radial symmetry, because there are infinite ways of slicing their body from top to bottom, all of which divide it into mirror images (like cutting a round Victoria sponge in half – sorry, I think I must be hungry).

The new head-to-tail axis of the bilaterians was key to their success. The invention of a head and a tail went hand in hand with the expansion of the number of Hox genes (whose job is to tell cells where they are on this new head-to-tail axis) and a flurry of related inventions and sophistications. Having a head gave the first bilaterians a more active, intentional, forward-facing life, exploring their environment, looking for mates, hunting for prey or avoiding predators. The sense organs – eyes most obviously – became grouped together at the front end, along with a gathering of nerves to form a brain. The mouth is also at the front, and the anus, another bilaterian novelty (and a good one!), is at the back. The more purposeful way of moving also required the invention of stiffening body cavities (coeloms which are pumped up with water and used as a hydrostatic skeleton); in the chordates, there is a backbone to which muscles can attach; and in several groups of larger animals there is a circulatory system to transport oxygen to the tissues of the more active body.

Several of these new bilaterian inventions are built from another unique bilaterian character – a new kind of tissue. To a good approximation, the non-bilaterian animals have just two layers of cells: the skin on the outside (called the ectoderm) and the gut on the inside (the endoderm). The bilaterians have evolved a third layer that lies between the skin and gut called the mesoderm. The mesoderm is the raw material for making new body parts that was not available to the jellyfish, sponges and friends. We bilaterians use it to build such things as muscles, skeleton, cartilage, tendons, blood, lymphatic system, heart, kidneys and even parts of our gonads. The potential for devising new body parts that was unlocked by the invention of the mesoderm was like the injection of new flavours into European cooking that arrived with plants from the new world (peppers, tomatoes, potatoes, chocolate, maize, beans, squashes and vanilla).

The Deuterostome Ancestor, 550 Million Years Ago

Urbilateria bequeathed to us (and to all of the bilaterian animals) a treasure chest of new characteristics. But the generosity of our urbilaterian ancestor, it has to be admitted, has not been matched by the ancestor we meet after our next rather dainty step up the tree. We are going, briefly, to meet the common ancestor of the deuterostomes – one of the two branches that emerge from Urbilateria (the other being the protostomes). The deuterostome branch leads on one side to the starfish and sea urchins (also to the hemichordates and, according to me, to the xenacoelomorphs) and on the other side to our own chordate branch, which we are going to follow shortly.

Work done in my lab in the past couple of years tells us that only a few million years separated Urbilateria from the first of the deuterostomes.[16] The brevity of this transition can be seen on the animal tree of life as a very short branch. To take this step from first bilaterian to first deuterostome, we barely move any higher on our human timescale, little more than a whisker, and this tiny span of time probably accounts for the paucity of new characteristics that distinguish deuterostomes from the other bilaterians.

Deuterostomes and protostomes, you may remember, are named for the different fates of the initial opening in an early embryo's gut.[17] This becomes an anus in the deuterostomes and a mouth in the protostomes. Textbooks tell us that protostomes and deuterostomes differ from each other in other ways also visible in their early embryonic development. The patterns their cells make as they first divide are supposedly different; in protostomes the cells are arranged in a spiral, and in deuterostomes the arrangement is more regular and is referred to as 'radial'. And the way in which the mesoderm is formed is also supposed to be fundamentally different in protostomes and deuterostomes.

In the past couple of decades, all three of the characters that supposedly distinguish deuterostomes from protostomes have turned out not to be unique to deuterostomes. Perhaps most tellingly, they are all also found in the arrow worms once considered to be bona fide deuterostomes but which we now know to be protostomes. We might therefore ask whether there is in fact anything that *is* unique to the deuterostomes. To think about it another way, is there any useful new character that the deuterostome ancestor has bequeathed to its descendants?

The one unique bit of morphology that the deuterostomes do seem to share is a series of gill slits, which at their most basic are simply holes piercing their throats (or, more properly, their pharynxes). Gill slits are used to expel excess water swallowed when feeding, and these are formed in deuterostome embryos when the cells of the pharynx (which is part of the gut) poke outwards to connect with the cells of the skin to make a tunnel to the outside. These pores or slits are easy to see in the gills of fishes, and, if you are ever lucky enough to meet one, you can also see them arranged along the neck of a hemichordate worm. Some deuterostomes have lost their gill slits – starfish, sea urchins and relatives don't have them, nor do terrestrial four-legged vertebrates. While you will search the mirror in vain for your own gill slits, they have nevertheless left important traces inside you, and we will come back to see their fate at the next stage of our climb up the tree of life where we meet the first chordate.

20
The Road to Mammals

IN 1911, CHARLES Doolittle Walcott described a new species, *Pikaia gracilens*, from his collection of Middle Cambrian Burgess Shale fossils. Walcott linked *Pikaia* to the annelid worms (earthworms and lugworms and relatives), but its fossils have long since been reinterpreted. *Pikaia* is now recognised as one of the very oldest known members of our own phylum of chordates.[1] It is undoubtedly fish-like, and Walcott recognised its likely swimming abilities: 'The study of a number of specimens of the posterior portion of the body leads me to think that it may have been flattened and thus been a much more effective aid in swimming.'[2]

The Chordate Ancestor, 520 Million Years Ago

Chordates are most obviously distinguished from the other living deuterostomes by an organ called the notochord, which gives chordates their name. The notochord, at its simplest, is a stiff (but bendy) rod lying closer to the back of the animal than to its belly and running the length of the body, from head to tail. *Pikaia* has one,[3] and it is easy to find your own equivalent because it has, over time and with the addition of cartilage and bone, become your spine. The notochord's most important job is to be a strong but flexible scaffold against which our muscles can work; muscles are attached to the notochord and can pull against it to cause the animal's body to bend (think of a fish bending side to side as it swims). Imagine removing a fish's backbone; the poor animal would, when its muscles contracted, fold up never to uncurl.

In addition to a notochord, chordates have blocks of muscle running along the body (these muscle blocks can be seen in the chunky flakes of flesh in a cooked cod); they have a tail that extends beyond the main part of the body and beyond the anus; they have a little organ in their pharynx that produces iodine-rich mucus (and which has transformed into the thyroid gland in our own necks); and they have a nerve cord (separate from the notochord) that has the form of a tube.

The position of the nerve cord is the clue to perhaps the most extraordinary feature of the chordates – we are all upside down. While all the non-chordates have a nerve cord that runs along their belly (and a gut that runs along the back side of their bodies), we chordates have the reverse – a dorsal nerve cord (inside your backbone) and a ventral gut. This seemingly drastic flip – the so-called 'axis-inversion' – could have been achieved fairly easily: one of our fish-like ancestors simply had to begin to swim upside down. What this means is that, from the point of view of an insect or earthworm, fishes swim upside down and we humans spend our entire lives walking backwards.*

The Vertebrate Ancestor, 440 Million Years Ago

The rate of change was very rapid around the beginning of the Cambrian, and the distance between successive ancestors – bilaterian, deuterostome, chordate, vertebrate – is short, so the progress on our human timescale has taken us barely beyond the middle of the chin. The first vertebrates appear after a mass extinction event that killed off more than 80% of marine species and which marks the end of the Ordovician period about 444 million years ago. It is in the warm seas of the Devonian period, starting about 420 million years ago, that the diversity of vertebrates really begins to take off. The flourishing of marine vertebrates has given this

* Because humans walk upright, what was once our downward-facing belly now faces forwards.

part of earth's history the unofficial name of 'the age of the fishes'. We have finally taken a noticeable step upwards to land on the upper lip of our timescale.

The glut of Devonian fossils stems in part from one special vertebrate invention. Mineralised hard parts – teeth and scales and bone – are durable enough to make complete and beautiful fossils. The many different characters we can discover in these early vertebrate fossils allow us to follow the accumulation of various vertebrate novelties over time: pairs of fins at front and back of the body (which have become our arms and legs); nostrils; the inner ear. Probably most important of all was the evolution of a head with a skull protecting an enlarged brain and, in the ancestor of what is now by far the most successful branch of vertebrates, jaws.* Contributing to this explosion of new characters being tried out by the Devonian fishes are two important events. The first is the invention of an entirely new type of tissue, and the second is a huge injection of new genes – the raw material of natural selection.

We have witnessed the impact of a new kind of tissue in the form of the bilaterian mesoderm. In addition to the three principal tissue layers that all the bilaterians share, we vertebrates have added a fourth called the neural crest.[4] Neural crest cells – both strange in their behaviour and rather wonderful in their effects – develop in a vertebrate embryo on the edges of the tube that is destined to become the nervous system. What happens after their first appearance is truly bizarre; the neural crest cells decide to move. Almost all movements of cells that happen when an embryo is growing are like the manoeuvres of a nineteenth-century army – coordinated mass movements of huge cohorts of cells. The cells of the neural crest behave very differently, more like Bedouin warriors: they prepare to move by separating themselves

* There were a great many varieties of jawless fishes, but just two tiny branches, the hagfish and the lampreys, survive today. Their greatest claim to fame may be their role in ending the reign of King Henry I of England, who, according to the chronicler Henry of Huntingdon, became unwell and died after eating 'a surfeit of lampreys'.

from the nervous tissue they come from, gathering themselves up to become amoeboid, as if packing up their tents. They now begin to move, individually, across the growing embryo, following invisible songlines, eventually coming to rest in a multitude of oases dotted along the developing animal. At this point they recognise each other again; they cluster together and start to cooperate to build up a host of different and important parts of the vertebrate body – major parts of the nervous system, most of the head and skeleton, the lenses of the eyes, pigmented cells in the skin, teeth, parts of the heart and more.

The evolution of our neural crest cells is considered by some to be one of the most significant events in the origin story of the vertebrates and the key to our eventual dominance of the land, sea and air. The influence of neural crest cells can be best understood from an example: we are going to revisit the gill slits that first appeared in the ancestral deuterostome.[5] In the simplest chordates, these are a series of simple pores piercing the pharynx; the pores have cartilage-like rods made up of stiffer cells that lie between them, helping to keep them open. In the early evolution of the vertebrates, the tissue between the gill pores became thickened and strengthened with muscle and cartilage; in vertebrate embryos today, these thickenings form a series of four bulges called the pharyngeal arches that are made from cells (muscle and bone) from the mesoderm. These can be easily seen just behind the head of an embryonic fish or chicken, or a four-week-old human embryo. The pharyngeal arches are one of the most important destinations for the streams of neural crest cells and they have gone on to make an outsize contribution to the formation of our heads.

The first contribution of the pharyngeal arches and associated neural crest cells came early in vertebrate evolution: the oldest vertebrates had a simple round mouth surrounded by teeth (and this kind of mouth still exists today in the jawless lampreys and hagfish). In our branch of vertebrates (the gnathostomes or 'jaw mouths'), the pair of pharyngeal arches closest to the mouth were modified over time to make the first jaws.[6] The contributions of the pharyngeal arches continues later: when aquatic fish became land-living tetrapods, the remaining arches were no longer needed to support gill openings and were co-opted for other purposes; those farthest back from the head, for example, moved inside and now make your parathyroid glands (which regulate the calcium in your blood) as well as the hyoid bone in your neck just below your jaw. The neural crest cells also make other big contributions to our complicated heads, forming the bones of our foreheads, the lenses of our eyes, the tiny bones in our middle ears, the vagus, trigeminal and facial nerves and the bits of our noses that allow us to detect odours.

Alongside the appearance of the neural crest (and possibly even partially responsible for its appearance), our jawed vertebrate ancestors were blessed with an enormous injection of new genes. These appeared all at once and as if by magic by the simple trick of duplicating the entire set of chromosomes. Indeed, this seems to have happened twice, effectively quadrupling the pieces in the Lego set that our vertebrate ancestors had to play with. The first hint of this double duplication came right back in the 1980s, when the vertebrate Hox genes were first discovered – where fruit flies have a single set of Hox genes, mice and humans were found to have four sets. Evidence for the double duplication can be found only in jawed vertebrates (the jawless lampreys and hagfishes branched off before the second duplication happened).[7] It has proved very tempting to conclude that many vertebrate innovations – neural crest, head, bony skeleton, jaws, fins and so on – were enabled by these injections of new genes. The success of the Devonian fishes must lie, at least in part, in this generous genetic inheritance.

The Tetrapod Ancestor, 390 Million Years Ago

Of all the directions taken by the many branches of Devonian fishes, it is hard to deny that the one with the most impact was the move from water onto dry land. This was a gradual process, as the body of a fish slowly changed, in fits and starts, into a form that could survive and then thrive in an utterly different world. The first evidence for the transition appears in the middle of the Devonian period, around 390 million years ago, with the appearance of fossilised trackways in the mud of a shallow lake.[8] These were made by an animal, still largely aquatic, whose fins were becoming leg-like, but which still dragged a fish-like tail behind it. The first fossils of an animal with a body that was intermediate between a lobe-finned fish (like a coelacanth) and a salamander-like early tetrapod appear in the Late Devonian around 375 million years ago.[9] We have reached the tip of the nose on our human timescale.

For a time in the early 2000s, I worked in an office next door to a pioneer in the study of this transition: Professor Jenny Clack. She was a brilliant scientist and a remarkable (expedition-leading, motorbike-riding, book-writing) person. She seemed to me to be rather shy (when I introduced myself, she said hello but failed to tell me who she was, and the conversation was at an end), but close colleagues and friends remark on her sense of humour. Clack's big contribution came soon after she began her career at the Museum of Zoology in Cambridge when she came across some partial Devonian 'fish' remains lying ignored in a drawer in the nearby Sedgwick Museum. These had been collected in Greenland in the 1960s, and it took Clack to recognise that they were not the fossils of just any fish but of a fish on its way to growing legs; they prompted her to lead new collecting trips to the spot the originals came from, and this work turned up many more specimens of this fish with legs (called *Acanthostega*) and its relatives.[10] The legs had counterparts of all the bones in our own limbs – humerus, radius and ulna; femur, fibula and tibia – as well as fingers and toes. The few specimens known before Clack began her work had been imagined as resembling a salamander or newt. Clack's work showed a much more interesting animal. All living tetrapods have five fingers or toes, and the rare exceptions, like the one-toed feet of horses and the three-fingered wings of birds, always have fewer, having lost digits along their evolutionary journey, but never more than five. *Acanthostega* and relatives, however, have seven or eight digits.[11] And the tiny bony parts of *Acanthostega*'s ear, previously interpreted as amphibian-like, seem, in these tetrapod relatives, to be much better designed for hearing in water than in air. They also had pretty decent fish-like tails. These animals, in short, were really as much water-living fish as they were land-living tetrapods, and Clack's chance discovery in a musty museum drawer would lead to the discovery of a wonderful series of snapshots of our ancestor's hesitant move from swimming to walking.[12]

Our ancestors were not the first members of the tree of life to move onto land; the plants and arthropods had got there first. But the demands of living on land, surrounded not by water but by air, required a rapid burst of invention among tetrapods: to avoid desiccation both of yourself and your eggs; to find a new way to breathe (lungs not gills); to find a new means of locomotion; and to support your full weight without the buoyancy of water. Your tongue is one of the many evolutionary inventions sparked by this move. Fish suck in bits of food by suddenly opening their mouths and producing a current, but our pioneering ancestors could not perform this trick out of the water. The pharyngeal arches, no longer required as supports for gills, were repurposed as supports for a tongue, which could manipulate food into and around the mouth. Not all of these innovations needed to happen at once (one can always pop back into the water to avoid desiccation and to lay eggs), and many changes will not have left traces we can detect in fossils, making it difficult to reconstruct the exact order of their appearances.

The transition from water to land and air and the evolution of four legs is one of the most famous and deeply studied events in all of evolutionary biology. All land-living vertebrates come from these Devonian ancestors. Three separate tetrapod branches have survived to populate the land today: amphibians (frogs, salamanders, newts and the little-known, legless caecilians); reptiles (including the feathered dinosaurs we call birds); and our branch, the mammals. The oldest fossil representative of this group of fully fledged tetrapods is a small, slender, salamander-like creature called *Westlothiana* (discovered in the limestone of the East Kirkton quarry, West Lothian, Scotland). *Westlothiana* lived 338 million years ago in the Carboniferous period, appearing at the end of a puzzling, 15-million-year-long part of earth's history called Romer's Gap, in which tetrapod fossils are mysteriously absent. Romer's Gap was eventually plugged by Clack and colleagues in a 2017 paper that described several new fossil tetrapods from Scotland.[13]

The Amniote Ancestor, 320 Million Years Ago

We are going to leave behind the amphibians to follow the other tetrapod branch which leads to the common ancestor of the mammals and reptiles; these two constitute a group called the amniotes. The amniote ancestor lived approximately 320 million years ago, and we have arrived just below the bridge of the nose on our human timescale. The amniote ancestor had evolved an armoury of characteristics that allowed it to venture away from the perimeter of a pond or the dampness of a rainforest, and these innovations, by enabling its descendants to survive in extremely arid environments, opened up countless new opportunities. The skin of amniotes has become thickened and waterproofed, so they are much more resistant to desiccation; their kidneys and large intestines have adapted to reabsorb the water that in a frog would be carelessly lost with excretion. Perhaps most important of all, amniotes have done away with the two-stage life cycle seen in the amphibians: a water-based larval stage that metamorphoses into a morphologically distinct, land-living adult stage. Amniotes instead hatch from their eggs already with the form of a miniature adult. The egg has a protective shell and 'amniotic' membranes, which let the embryo absorb oxygen without losing water by evaporation (and which give the amniotes their name). From the amniote ancestor we are now going to follow the branch that leads to the mammals.

21

The End of the Journey

For many people, mammals are what come to mind when asked to picture an animal. Mammals are very diverse and now live in almost every ecosystem on earth. They can be found, often in huge numbers, from the tropics to the poles; on land, river, sea and in the air; mountain and valley; prairie, forest, steppe and in our houses both as welcome guests and as pests. They range in size from the very tiny Etruscan shrew to the absolutely enormous – we are lucky to share the planet with the largest animal that has ever lived – the blue whale. They may be placid (a manatee or a sloth) or fierce (a Tasmanian devil or honey badger), cute (pick your own but for me it might be a slow loris with its sad eyes and soft fingers) or less lovely (a naked mole rat?). They are predators, omnivores, scavengers, herbivores, piscivores, insectivores, sanguinivores and pretty much everything elsivores.

The Mammal Ancestor, 180 Million Years Ago

While today we can immediately recognise a mammal as a mammal, for tens of millions of years after the mammal branch separated from the reptiles, our ancestors had little in common with a cow, dog, mouse or monkey. The very first of the mammal lineage looked like reptiles, and the mammalian characteristics we are familiar with accumulated only very gradually. The end point of all this accumulation of mammalian characters was the mammal common ancestor, which may have lived as long ago as

180 million years. The common ancestor is survived by two groups. The first, the monotreme branch, leads to just five species (the duck-billed platypus and four species of echidna), all of which still lay amniote eggs. The second mammalian branch (called Theria from the Greek for 'wild beast') leads to the hugely successful and diverse host of mammals that can be recognised by their abandonment of the amniote egg for a warm, wet, protective and nourishing womb. Theria contains both marsupials (which give birth to tiny, essentially foetal babies which climb into their mother's pouch to mature into a baby kangaroo or opossum or koala or thylacine) and eutherians – better known as the placental mammals, whose much longer pregnancies produce large, well-developed babies.

The mammalian ancestor must, of course, have had a hairy body and the ability to lactate – things we and kangaroos and the duck-billed platypuses all share. Most other unique characters that define the mammals are internal and so less easy to see: the ability of our young to suck and swallow milk depended on the evolution of bones with flexible joints at the back of the throat; our keen sense of hearing depended on the evolution of three tiny bones – malleus, incus and stapes (hammer, anvil and stirrup) – connected to each other in a series in our middle ear. Our broad diets and opportunistic omnivory were enabled by the evolution of teeth of varied shapes, each with different roles – think of your own incisors, canines, premolars and molars for cutting, tearing, crushing and grinding.

We have a natural interest in knowing the order and timing of the events on this branch that gradually added up to make the first mammal: what came first – suckling? fur? different-shaped teeth? These events happened somewhere along the 140-million-year branch that separates the ancestor of the living mammals (an entity we can picture if we compare platypuses, kangaroos and cows) from the much older amniote common ancestor (which we can know by comparing mammals, reptiles and birds). But between these two ancestors there is a yawning gap – a long, bare branch from which no *living* branches emerge, and the result is a 140-million-year *terra incognita* where our time machine cannot travel.

In truth, this branch isn't bare – there are countless side branches of now extinct species that, if we follow them through the fossil record, are more and more closely related to the mammals that survive today. Examining these fossilised remains lets us trace some of the changes that turned a reptilian amniote ancestor into a familiar mammal.[1] The skull of the amniote ancestor had two sets of paired holes (left and right) which allowed for the attachment of powerful jaw muscles. Both pairs can still be found in the skulls of reptiles and birds, but in mammals one of these pairs has closed over; this fused hole is one of the first mammalian features to appear. The fused holes make a solid arch below the mammalian eyes, which you can feel in your own skull in the form of your cheekbone. The first evidence of this solid arch can be found in *Asaphestera platyris* and in the huge, herbivorous *Cotylorhynchus* – which was almost 7 metres long and lived 320 million years ago.[2] Then, 295 million years ago, and one step closer to modern mammals, there is *Dimetrodon* – still huge, and giving the first evidence of multiple tooth types (*Dimetrodon* means 'two measures of teeth'). In *Thrinaxodon*, which lived 250 million years ago, we can find the beginnings of a flexible hyoid bone in the throat, and *Microdocodon*, which walked the earth 165 million years ago, developed this further. These hyoid bones provide us with the earliest evidence of suckling, and this in turn implies the existence of mammary glands. The 165-million-year-old fossil of *Megaconus* contains ancient evidence of fur.[3] And finally, in the rat-sized *Liaoconodon*, we can see an early stage, some 120 million years old, of the jaw bones that were changing their shape and position as they morphed into the tiny bones of the middle ear.[4]

The discoveries of various weird and wonderful proto-mammals are rewriting our own ancient history. The traditional story is that early mammals existed as tiny shrew-like creatures, living in burrows and only emerging at night, existing nervously and precariously in the shadow of the dinosaurs and flourishing only after the asteroid wiped out the competition. There is some truth in this story, but fossils show that the early mammals could be much larger than this picture suggests and indeed much more diverse. One fossil species called *Repenomamus giganticus* lived about 125 million years ago in the Early Cretaceous (contemporary with the dinosaurs *Iguanodon* and *Baryonyx*). *Repenomamus* weighed about 12 kilograms (perhaps *giganticus* is a bit of an exaggeration), and, in a Cretaceous version of 'man bites dog', one of its fossils has been discovered with baby dinosaur bones in its stomach.[5] Flying, swimming, tree-climbing, digging and hunting species of early mammals are all known from the time of the dinosaurs.

The Placental Mammal Ancestor, 90 Million Years Ago

In the past three decades, our view of the relationships *between* placental mammals has been radically upended by molecular data. Older trees, which were based on comparisons of mammalian morphology, had been misled by the problems of convergent evolution. Classic examples of errors include a branch of insectivores containing moles, golden moles, hedgehogs and tenrecs and a branch containing the ant- and termite-eating mammals – armadillos, anteaters, aardvarks and pangolins – which all have long snouts, long tongues and powerful, clawed front limbs.[6] DNA sequencing has revealed that neither of these groups is real. Animals such as bats, whales and manatees that have drastically reshaped their bodies have also confounded the mammal tree, their unusual bodies obscuring their relationships to other mammals.

During the Cretaceous, the supercontinent of Pangea broke up to form the continents we know today. These huge land masses, separated by vast if shallow seas, became nurseries for many new kinds of mammals. The tree of placental mammals we know today has three great boughs that are defined by their continents of origin.[7] The South American group is called Xenarthra ('strange joints' for the unusual way that their vertebrae fit together), which consists of anteaters, tree sloths and armadillos. The geographic origins of the two other larger groups may be more immediately clear from their names: Afrotheria ('African mammals') is a hyper-diverse group that includes elephants, elephant shrews, aardvarks, hyraxes, manatees and dugongs; and Boreoeutheria ('northern mammals') includes hedgehogs, rodents, pangolins, moles, bats, wolves and monkeys.

The evidence for convergent evolution jumps right out from this new tree; ant-eating animals can be discovered scattered amongst all three branches – armadillos and anteaters in South American Xenarthra, aardvarks in Afrotheria and pangolins in Boreoeutheria. The moles and hedgehogs are boreoeutherians; the golden moles and tenrecs are afrotherians. The bats, whales and dolphins are all northern mammals. The whales are nestled amongst the even-toed ungulates – the giraffes, deer, camels, pigs and (closest of all to the whales) the hippos. Our own mammalian order of primates, finally, is part of a division of Boreoeutheria called Euarchontoglires – an unfriendly name that is a mash-up of *euarchonta* ('true rulers' – a nod to the supremacy of the primates) and *glires*, which refers to dormice (and by extension to the rodents). Within Euarchontoglires, the closest relatives of the primates are the tree shrews, rodents, rabbits and flying lemurs.

The Primate Ancestor, 75 Million Years Ago

This long and winding path up the tree of life has brought us to within 75 million years of our final destination, to the ancestor of a group immodestly named by Linnaeus for its primacy – the primates. As Winston Churchill was fond of saying, history is written by the victors. We primates all share a rounded skull that protects a brain that is quite large for our body size; our eyes are forward-facing, set in a more or less flattened face; we have fingers and toes that end in a flat nail, not a claw; our big toes and thumbs are separated from the other digits of the foot or hand and so are typically opposable and can be used to grasp objects (branches or tools or weapons); we have fingerprints and toeprints; and, unlike other mammals, our molar teeth have rounded tops.

The primate branch contains groups which naturally seem familiar and, let's admit it, especially attractive and charismatic. Our most distant primate relatives are the lemurs of Madagascar and the wide-eyed lorises and bush babies; closer to humans are the monkeys of the new world (spider monkeys, tamarins and capuchins); and closer still are the old world monkeys (macaques, baboons, langurs and others). The ancestor of our own group, the apes (recognisable by the lack of a tail), lived roughly 18 million years ago, high in the hairline of our human timescale.

The Ape Ancestor, 20 Million Years Ago

Perhaps there was a careful cost–benefit analysis conducted by the process of evolution at the time, but for me the defining characteristic of the apes – the lack of a tail – has always been a bit of a disappointment. If we brought back the tail, we would all have to buy new trousers, and chairs might need redesigning, but I wish that I still had mine – I imagine a magnificent bushy one like a ring-tailed lemur's. This loss has recently been attributed to a change to a gene called *Brachyury*. *Brachyury* was first studied in 1927 by the extraordinary Russian émigré Nadiya Dobrovolska-Zavadska, who, following the defeat of the White Russian army (into which she had been conscripted as a medic), had fled to Paris and begun an entirely new career at the Pasteur Institute as a geneticist.[8] By irradiating the testicles of countless mice with X-rays, Dobrovolska-Zavadska was able to introduce a host of new mutations, one of which affected the growth of the mice's tails. The resulting stumpy-tailed mice gave the mutated gene its name – *Brachyury* means 'short tail'; the stumpy tail of the Manx cat (from the Isle of Man) has also been traced to a mutation in *Brachyury*.

Brachyury is just one of several dozen genes that are known to have roles in making the mammalian tail during embryogenesis, but it seems by far the most likely culprit in the loss of our tails. In every species of ape there is a piece of foreign DNA (a virus-like fragment) that can be found jammed into the middle of their *Brachyury* gene; and the effect of this insult to the DNA is the lopping off of a chunk of the Brachyury protein that is encoded by this gene.[9] While we may know the genetic cause of the loss of the tail, it is all but impossible to know for certain what the benefit of losing a tail was. It is often suggested that ditching the tail made it easier to walk on two legs, but I'm not so sure – tails are often rather helpful for balancing, and besides, the other apes do not generally walk two-legged. A more prosaic explanation is that, just like the Manx cat, our ancestor lost its tail simply because of an unlucky break; a tolerable mutation occurred that spread through a small population of apes which happened to become successful for unrelated reasons – despite, rather than thanks to, the loss of their appendage. Loss of 'advanced' characters like a tail, whether by accident or design, has become one of the most important themes of our long journey to humans.

Within the tailless apes we find five familiar branches: the small, long-armed, whooping gibbons (most distant and most different from us), the orangutans (slightly closer), and the gorillas, chimpanzees and humans, whose ancestors separated from one another in quick succession between 10 and 8 million years ago. The short space between branches meant that it took a bit of working out, but we now know that our closest living kin on the entire tree of life (closer than gorillas by the slenderest of whiskers) are the two species of chimpanzee – chimps and bonobos. Not one species of non-human ape can be said to be thriving. Lumping together every individual in all twenty-seven species (twenty gibbons, three orangutans, two gorillas and two chimpanzees) gives a total of around a million non-human apes, including critically endangered populations of 7,000 Sumatran orangutans, 5,000 eastern gorillas and as few as twenty-two Hainan black-crested gibbons.[10] For hundreds of thousands of years, *Homo sapiens* existed as just another fairly successful species of ape with perhaps 100,000 to 1 million individuals; today, we outnumber all the other apes combined by roughly 8,000 to 1.

The Human Ancestor, 8 Million Years Ago

In the 8 million years since our human branch parted from the chimpanzee branch, the species on our branch have come down from the trees and begun to walk upright, run, swim, talk, make tools and art and live in large (and now vast) societies. Our bodies have changed too: less hair, a chin, permanent breasts, a bigger brain, straighter legs, shorter arms, larger skull.

It is inevitable, perhaps, that we are more alert to the differences between us and our nearest animal neighbours than to the similarities, and that the events of the most recent 8 million years of our own history fascinate us more than the previous 4 billion. What makes us human, why are we as we are – gregarious, talkative, inventive and, far above all else, intelligent? Scientists are looking for the answers in the differences between human and ape DNA; and there is a parallel search across the planet for the series of extinct ancestors and near relatives that once bushed off the branch that leads from the human-chimp ancestor to *Homo sapiens*.

This final branch on our human-centric tree of life, whose closest relatives are the chimps and whose only living leaf is *Homo sapiens*, encompasses a larger group called the hominins – all but *Homo sapiens* now extinct. The rest of them knowable only through their fossils. A few of these extinct hominin species must constitute a chain of direct ancestors that, in a million imperceptibly tiny steps, were transforming into us. To add confusion to the interpretation of the bashed-up, squashed and always partial remains of our ancestors, there is a second category of extinct hominin species that are not our direct ancestors but lie instead on side branches – distant cousins, not great-great-grandparents but still able to tell us about the arrival of new human characteristics.

The oldest hominins are the most confusing of all – all but indistinguishable from the first members of the chimp lineage. With time, however, inklings of the new characters that were to forge a modern human body could be recognised. The oldest hominin fossil we have found so far, *Sahelanthropus tchadensis* ('Saharan hominin from Chad'), lived about 7 million years ago.[11] *Sahelanthropus* was in many ways chimp-like, slight in stature and with a small brain, but its teeth reveal it as a human relative. *Sahelanthropus* has small, human-sized canine teeth, very different to the long and chunky fangs of chimps; like humans it had lost the gap in its upper jaw (between incisor and upper canine) that once accommodated the large lower canines; and like humans its small upper canines were worn down at their tips rather than honed to sharpness on their edges like a chimp's. *Sahelanthropus* was also probably already walking on two legs, at least some of the time.[12] The hole in the base of its skull is close to the centre, as in humans, helping to balance the head atop the spine while in an upright position.

There are other fossil hominins almost as old:[13] *Orrorin tugenensis* (found in Kenya and roughly 6 million years old) had *Homo*-like leg bones and also walked upright, but a robust humerus shows it had muscular arms, suggesting time spent climbing trees as well. *Ardipithecus* ('ground-living ape', found in Ethiopia and 4.4 million years old) had evolved a broad, human-like pelvis that would have made walking more efficient – no more swaying from side to side like a sailor in a storm. But *Ardipithecus* also had curved, branch-grasping fingers and powerful legs and buttocks made for climbing. When it came to moving around, our ancestors had become multitaskers.

Clues about the environment our ancestors lived in can be teased out of the material lying around their skeletons. Fish, otter and crocodile remains found alongside *Sahelanthropus* give evidence of a body of water; horse, giraffe and elephant bones and teeth tell us about a nearby savannah (perhaps these were animals that had come to the water to drink); monkey bones suggest a ribbon of trees fringing the water; and sand grains are evidence that this water was surrounded by desert.[14] Clues surrounding *Ardipithecus* show it lived in a forest alongside monkeys and parrots and peacocks, and probably ate the figs and palm fruits whose seeds can be found nearby.

Four million years have passed, and I doubt that if we time-machined back to meet *Ardipithecus* in the flesh we'd recognise these small, hairy, small-brained creatures as close kin; sure, they're able to walk, but there is precious little more that is recognisably human to show for 3 million years of evolution. The next step, however, brings something really profound. It is impossible to know for sure what prompted the change, but around 4 million years ago a new group of hominin species began making tools. These species, responsible for the footprints fossilised in the ash at Laetoli, have been given their own genus, *Australopithecus* ('southern African ape').

Some *Australopithecus* tools (when made from long-lasting stone rather than perishable bone or wood or hide) have survived close to *Australopithecus* remains, and nearby animal bones bear scratches from meat-cutting tools. But the best evidence for tool use comes from the fossils themselves. The bones of the wrist and hand have taken on a form that would have given australopithecines excellent handling capabilities for manipulating small objects, as well as a precision grip, perhaps to keep these objects steady during carving. The changes also made possible a power squeeze of the whole hand (different to the solid lock of fingers used to swing from branches), suggesting the ability to wield loose, large objects, such as a stone axe. It seems plausible that this new, delicate hand was only possible because the arms were no longer being used for walking.

Our own genus of *Homo* finally makes its appearance in the fossil record about 2 million years ago.[15] Our human timescale, which we have used to measure billions of years, has become almost worthless, like trying to find your way round a village using a map of the world; the rest of human history (and we have still to arrive at *Homo sapiens*, remember) fits comfortably into the last millimetre of skin on the top of our human timescale's head. Fossil humans (if we use this name for all members of *Homo*) can be recognised for a whole host of new features: smaller teeth, a smaller and less out-jutting jaw, a nose that sticks out from the face, longer legs, a bigger brain. The first evidence for these features (and so for *Homo*) comes in maddeningly incomplete fossil skeletons. It is unclear how many species there were, whether the big brain case found over here comes from the same species as the long leg bones found elsewhere, or even whether the same characters may have been evolving repeatedly (convergently) in different groups.[16]

It is only with the arrival of fairly complete fossils of a species called *Homo erectus* ('upright man', roughly 1.5 million years old) that we can be fully confident we are gazing into the eye sockets of a member of our own genus. *Homo erectus* first evolved in Africa perhaps 2 million years ago but quickly spread across the globe – the first fossils, discovered in the 1890s, came from Java, Indonesia, and others have been discovered in China and Georgia. Although *Homo erectus* possessed all the human characters listed above and could grow as tall as modern humans, we would still immediately recognise them as different. They were more powerfully built than we are, they had thicker skulls and smaller brains, chunkier jaws, a jutting brow ridge and no chin. There are other variations around the central theme of *Homo* known from the past 2 million years: *Homo antecessor*, *Homo heidelbergensis*, *Homo naledi*, *Homo habilis* and, depending on the interpretation of certain fossils, several others.

We are finally arriving at the terminus of our long journey. The last of all the splits on the tree separates *Homo sapiens* from *Homo neanderthalensis* – the Neanderthals. These two sister species diverged from a common ancestor that lived in Africa about 800,000 years ago, and they lived through a time of global change and ice ages.[17] The discovery of our closest cousins had been anticipated by Ernst Haeckel, who included them in his tree of primates with the placeholder species name of *Homo stupidus*. The name we ended up using derives from 'Neandertal', the name of the German valley in which the first described fossil was found (*tal* means 'valley' and *Ne-ander*, coincidentally, means 'new man'). They were shorter than modern humans, with a stocky, solid body and short arms and legs; a heavy brow ridge; big, slightly jutting out teeth; and flatter noses.[18] We struggle to imagine them as anything other than brutish troglodytes, but their brains were slightly larger than ours, they used tools and made ornaments, cooked and kept warm with fire. The hyoid bones from their throats are just like ours, telling us that they could almost certainly speak. Some of our *Homo sapiens* ancestors certainly didn't find them too unappealing – the snippets of Neanderthal DNA we still carry in our own genomes are witness to children produced in ancient interspecies love matches.

While the Neanderthals dominated in Europe and Asia, we *Homo sapiens* did our growing up in Africa before spreading out across the world roughly 60,000 years ago in waves of migration and conquest, eventually crossing the Bering Strait from east Asia to North America about 20,000 years ago and working our way down the west coast to South America. For tens of thousands of years, we lived alongside other species in our genus, but, eventually, we outcompeted and replaced them everywhere.

The domination of the planet by *Homo sapiens* might seem the inevitable consequence of our superior intelligence and adaptability, but in truth our existence today involves a good bit of luck. The slight variations that exist between the genes of modern humans from across the planet bear witness to a time eight or nine hundred thousand years ago when the genus *Homo* stood on the edge of disaster. The DNA records a time when the total population of ancient humans crashed suddenly from a healthy 100,000 individuals to a mere 1,000.[19] This tiny population seems to have limped on in this precarious state for more than 100,000 years before recovering. Today, we number more than 8 billion; the total mass of our single species is equal to ten times the total mass of *every individual of every single other species of wild land mammal combined* (although we are still outweighed by the mammals we farm).[20]

We have traced our own long, unpredictable and perilous path up the tree of life, witnessing the changes, additions and occasional subtractions of characteristics that have combined to make a human. Almost every species that the tree of life ever produced is now extinct, and we are one of the few lucky ones to have dodged the hail of bullets – changing climate, disease, predators, food scarcity, competition and bad luck. We have now arrived in the present – with you and me. This moment seems like it ought to be the summit of our climb, but there is an argument to be made that this is a false peak, that there is another summit just a few steps away that is higher by a whisper. In the next chapter we will discover that we can step out onto this slightly higher ground by looking inside ourselves, where we will find other tiny trees. Trees that describe the relationships between our cells and between our genes, and even the evolution of our culture.

22

The Trees Within Us

WHILE A TREE is an elegant way to represent the relationships between species, a tree can also depict the evolution of many different things that change over time. The two simple requirements are: a process equivalent to speciation (a single individual of one generation splitting to give more than one in the subsequent generation or at a later time); and the more or less faithful passing on of some form of malleable information. If these conditions hold then reading the characteristics of the 'species' at the tips of the branches will let us reconstruct this tree of relationships and see the history of their ancestors.

Possibly the most famous example of non-genetic evolution can be found in human culture. We can reconstruct trees relating languages, and individual words and the books that contain them. We can also use trees to represent the histories of other parts of culture: arts and crafts, customs, laws, religions, technology, recipes and designs. The signals from the pulsating skin of a cuttlefish are, as far as we can tell, a form of language that is hard-coded in the cuttlefish's genes rather than learned from other members of the species. It has evolved like any other characteristic of the cuttlefish, by changes in its DNA that have been handed down from parent to offspring. Human language (indeed all of human culture) is very different: while our genes bless us with the capacity to learn to speak, our language itself exists not in the sequences of nucleotides in our DNA but (in ways we do not understand) stored in the tangled web of neurons in our brains. As Richard Dawkins has argued, languages and ideas (memes) behave for all the world like genes, passed down the generations, sometimes changing, the best new versions prospering, the runts of the litter failing to thrive and fading away.[1] The modern version of a meme – videos of cats and of people falling over – is a sorry shadow of the original idea.

The parallels between language evolution and species evolution, although not perfect, are still striking: genes and nucleotides correspond to words and letters; DNA replication to learning from your parents; mutations to mistakes and neologisms; speciation to human migration and the splitting of populations. Both species and languages can go extinct (Darwin mentions a parrot seen by Alexander von Humboldt in South America that was 'the sole living creature that could speak a word of the language of a lost tribe'[2]), and both genomes and languages can merge to make a hybrid.

The same methods we use to build and to interpret the tree of life can be used to reconstruct lineage trees of languages to reveal their past history. Just as we can use homologous arms and wings and fins to reconstruct the limbs of the first tetrapods, different Indo-European languages have been compared to detect the homologous words that descend from a common ancestor: *dan* (Hittite), *dve* (Sanskrit), *dwa* (Polish), *dha* (Irish), *deux* (French) and *due* (Italian) are all homologs of the English word 'two'; and the relationships between these homologs allows us to time-travel back to their common ancestor to know a word that must have been spoken by a proto-Indo-European six to seven thousand years ago.[3]

Even some of the less tree-like aspects of language evolution have their counterparts in the problems of mergers and hybridisations that affect the tree of life. Most obviously, a counterpart of the hybridisation of species can be found in almost every sentence in this book: the mixture of English word origins *reveals the grafting* of Latin, French and Old Norse branches of the language tree onto the Germanic 'English' branch: 'reveals' comes from Latin via French; 'the' from Germanic; 'grafting' from Old Norse.*

* I swear on *On the Origin of Species* that I picked these three words at random.

The written word has often been used to represent the way that genetics and evolution work. We might be asked to imagine a book in a language that uses just four letters, A, C, G and T. Words in this language are all three letters long, and the 'meaning' of each word is an amino acid. Strings of words – sentences and paragraphs – correspond to whole proteins, and the whole genome's worth of genes corresponds to an entire book. We could go even further and, to explain how species evolve, imagine a series of scribes copying out the book in the equivalent of generations of a species. Each scribe would make occasional mistakes which would, in our simile, correspond to mutations which would be passed on when those scripts are copied in future.

The simile is so good, the parallels so close, that it is tempting to imagine we could use our ideas of species evolution to understand the tree of relationships between books. It turns out, naturally enough, that this idea has occurred to plenty of people, and the New Testament is an exemplary text for this purpose.

From the third century CE onwards, we have many versions of the Gospels. There are Greek and Latin versions alongside Gothic (as in the East Germanic Goths), Armenian, Ethiopian and Egyptian. The different copies can vary significantly; more than 100,000 textual variants (mutations, characters?) of different kinds (additions, substitutions, omissions of various types) had been catalogued by the nineteenth century. The material to reconstruct a genealogical tree of the many versions of the New Testament is there in abundance. The motivations to do so may seem familiar from the uses we have seen for the tree of life itself. At least in the nineteenth century, the aim first and foremost was 'to ascertain and restore, as far as possible, the original text as it came from the pens of the apostolic authors . . . to show not what they ought to have written, but what they actually did write'.[4] In biological terms, their purpose was to get as close as possible to the last universal common ancestor of the Gospels.

A tree of bibles can be put to other uses, which again map closely to the ways in which we use the tree of life. In the same way that a tree of genes revealed the dual origins of eukaryotes in a merger between bacteria and archaeans, a tree of biblical texts might reveal the origin of a new text thanks to a merging of two distinct older texts – perhaps one from Egypt and another from Constantinople. Identifying this merger takes us back in time to picture the texts that a long-dead scholar had in front of them as they sat down to produce a new version. We would even be able to date the moment of transcription – it must have happened before the appearance of the mixed version but later than the origin of both of the two sources. This dating of ancestors and this merging of branches are, of course, both familiar themes from the tree of life.

It is not just our external, cultural world that can be understood by thinking about trees; we can use a tree to understand our inner selves too. As you read this, you can picture yourself on the tiny *Homo sapiens* leaf, balanced at the very tip of a twig on the enormous tree of life. Though this looks like the natural end of our story, there are in truth two further steps to take. If your eyes were only acute enough, you would be able to look right inside yourself where you could find two more tiny trees secreted within your leaf on the tree of life.

The first of these tiny trees tells us about the relationships between the individual cells of your body, present now in their trillions but all arising from a single ancestor cell – the fertilised egg – that existed at the moment of your conception. The fertilised egg is the equivalent of LUCA, and the host of cells that exist in you today are the counterparts of the millions of species alive today that are LUCA's descendants. There is a second even smaller set of trees inside your cell nuclei, each of them describing the relationships between a little family of genes.

At the root of the first of our internal trees, the one showing the relationships between our cells, lies the fertilised egg cell. Soon after fertilisation this ancestor divided to make two daughter cells, just as the single species of LUCA divided to found the archaeal and eubacterial branches of life. Each of the two daughter cells divided in two again to make a total of four granddaughter cells, which in turn divided to make eight, and, through the magic of exponential growth (16, 32, 64, 128, 256, 512, 1024 . . .), forty-odd cell divisions later, at the moment of your birth, your cells numbered as many as 15 trillion.

Knowing how the cells of our body are related to one another, just like knowing the structure of the tree of life, is a key tool for understanding the processes that build our bodies during our growth in the womb. If we could know our cells' family tree, we could ask, for example, whether cells are related according to their functions or according to location in the body: is a muscle cell in your foot more closely related to a nerve cell in your foot or to a muscle cell in your jaw? We could also ask whether every human has the exact same cell lineage tree (which would tell us that human embryogenesis is very regimented) or whether yours differs from mine (which would suggest that human embryogenesis works using rules of thumb rather than a rigid protocol). Knowing how the different kinds of cell in your body are related to one another might reveal hidden connections between them; are the very different cells of your blood – oxygen-carrying red blood cells, bacteria-killing white blood cells, monocytes, neutrophils, eosinophils, basophils, and macrophages – all members of a single family of cells or do they have independent origins?

Working out the first animal cell lineage tree (work completed in 1983) earned Sir John Sulston a Nobel Prize. Sulston's celebrated laboratory animal is a species of tiny nematode worm called *Caenorhabditis elegans* (almost always known simply as *C. elegans*). Sulston's work was made possible because *C. elegans* is tiny (a millimetre long), it is see-through and its embryonic development is fast, the entire business taking just over thirteen hours. Moreover, each worm has just 671 cells, 113 of which (or 111 in the male), having served some function and being no longer required, undergo a sinister-sounding process called 'programmed cell death'. These convenient attributes meant that all the cell divisions of embryogenesis could be followed in real time by staring down a microscope.[5] Most conveniently, the cells of every single worm of this species follow the exact same path. The same cells reliably divide at the same time points and in the same places, and each one of the cells at the tips of the cell lineage tree's branches will end up producing, in every single worm, the same one type of larval cell, be it a neuronal cell or a muscle cell or a gut cell. This amazing consistency meant that Sulston didn't need to get everything down in a single pass. He could observe the exact same embryological processes again and again in countless identical embryos.

The nematode cell lineage tree was published over forty years ago and is still being used to this day. The insights it has generated have been many and various, but I will describe just two. The first, alluded to above in the context of a human, concerns the origins of different cell types. One obvious assumption might be that neuronal cells should be closely related to one another on one branch of the lineage tree, muscle cells on another and so on. What the *C. elegans* cell lineage tree revealed, however, is that, for this worm at least, this is not the case. Different branches of the worm's cell lineage tree are each able to produce multiple different cell types – muscle cells, neurons and gut cells. This surprising result is telling us something important about how worm embryos are really made, how development works. Distantly related branches of the worm's cell lineage tree follow the same carefully orchestrated set of cell divisions to produce the same sets of cells. It is as if they were all running the same bit of computer code.

The second insight gives us one tiny scintilla of evidence about how animals come to look different from one another. The modern version of Darwin's theory (*Origin of Species* with a twentieth-century appendix on genetics) explains how mutations in genes produce offspring that look different from each other, and that these differences are the raw material of natural selection. But how on earth does changing a nucleotide in a gene actually lead to a difference in what a worm (or a human) looks like? The general answer is that you need to change the genes that guide the building of their body – the genes that control embryogenesis.

In our case study, we are going to see how a change to the genes guiding nematode worm embryogenesis produced two closely related species of worm that look different from each other. The comparison was only possible thanks to the unchanging cell lineage of both worms, which allowed equivalent twigs on their cell lineage trees to be identified.

C. elegans is an unusual animal in many ways (tiny, transparent and so on), but one notable oddity is that there are sperm-producing males but no females; there is a second sex that lays eggs, but it is not a female but a hermaphrodite that produces both eggs and sperm. At some point in its embryogenesis, the cell lineage of a *C. elegans* hermaphrodite produces two cells named Z1aa and Z4pp, which lie on different branches of the lineage tree.[6] The task of these two cells is to persuade other cells to follow them as they move along the growing embryo, one of them forward the other backwards; the resulting congas of cells will later form two hollow tubes that eggs can pass through when being laid – the twin *C. elegans* vulva is described as being 'didelphic'.

The cell lineage tree of a closely related nematode called *Panagrolaimus* has also been followed, and the exact same Z1aa and Z4pp cells have been identified. But evolution has made one tiny change to the programme of embryogenesis in *Panagrolaimus* – the effect of which is to kill a single cell of the several hundred in the worm. The designated victim is Z4pp, which, instead of guiding the formation of the posterior vulva (as it does in *C. elegans*), falls on its sword to undergo programmed cell death. And with Z4pp dead, there is, of course, no cell to lead one of the two vulva-forming congas. The striking result of such a tiny change to a single cell is that *Panagrolaimus* has only the forward-pointing branch of the vulva; it becomes 'monodelphic'.

Sulston's method for working out the cell lineage of *C. elegans* was slow and laborious and only possible thanks to the unusual properties of the worm. Most animals have none of *C. elegans'* advantages: they have vastly more cells – even the first and smallest of the three larval stages of a tiny fruit fly has 50,000 cells, and a mouse has well over a billion (cells are really small!). On top of this, most animals are not transparent; and if you were interested in knowing the cell lineage of a mammal, our development takes place hidden inside our mother's womb. Finally, most animals do not have the metronomically consistent pattern of cell divisions that the nematode does. If we want to reconstruct cell lineage trees for a fruit fly or a mouse or a human so that we might use it to understand more about how our bodies are made, we are going to need a completely different approach, and there is one that seems like it might possibly work; it owes everything to the methods we have explored for constructing the tree of life.

The modern approach to constructing the tree of life that relates species uses molecules as a record of the past. Nucleotides in DNA and amino acids in proteins are a repository of all the mutations each species has inherited from the ancestors that lie in its past. It is this store of information that is now being used to study the relationships between cells in a single organism. Individual cells in a growing embryo can experience mutations. These are rare – just 1 of the 3 billion nucleotides is typically mutated in each cell division – and, just as with species, these changes to DNA are passed down to the cell's offspring. There are just enough of these rare mutations to provide the raw data we need to work out the relationships between cells.

The objective of the experiment is to read the accumulated mutations in all the cells of a newly hatched fruit fly or newborn mouse and to use these to build a cell lineage tree, but the experiment is really tough. The first hurdle is to find the tiny number of mutated nucleotides inside an individual cell (a needle in the 3 billion-nucleotide haystack of the genome). This problem is then multiplied enormously because, to learn something useful about the cell lineage, we must achieve this herculean task in anything from tens of thousands to billons of separate cells. The first tentative steps to knowing cell lineages of animals more complex than a tiny worm are only now being taken as labs around the world race to complete this ambitious experiment. When it succeeds, the insights it will give into the secrets of embryogenesis promise to be as transformative as Hooke's drawings of the insect world that followed the invention of the microscope.

The story we have told of cell lineage trees has shown that, while we have been preoccupied with the relationships between species on the tree of life, we can just as easily follow the evolution of smaller parts of biology: individual characters, cells and genes. While genes in particular generally evolve in step with the species they are a part of (we depend on this correspondence when building the tree of life), the evolutionary history of a gene does not always match perfectly with that of the species it inhabits. In these occasional discrepancies, interesting parts of the story of evolution lie waiting to be discovered.

The occasional schism between histories of gene evolution and of species evolution is most blatant in cases of horizontal gene transfer, such as occurred in the forging of the eukaryotes. This is an extreme example, but it is also a hint that it might be interesting to think about genes individually and separately from the organisms they inhabit. The evolutionary pasts of at least some genes turn out to be more eventful than simply going along for the ride in a single species.

The world around us is awash with colour. Much of the colour in nature has no special meaning, there by accident and not design: plants are green not to please the eye but simply because chlorophyll is green; rocks are grey or brown or yellow regardless of whether anyone is looking – they look the same on Mars and Pluto; and the sky is blue because blue photons scatter more easily than green or red ones do, and not to entrance a poet. But there are many colours in the natural world that *do* exist for a reason: they have been designed by evolution to catch the eye. The blue of a cornflower exists because it can be perceived by the eyes of a bee; the vivid colours of a poison arrow frog – dark reds and blacks, greens, blues, oranges and yellows – exist as a warning to would-be predators that have eyes able to see them; the camouflage of a stick insect helps it avoid the gaze of a chameleon or a bird; and the exuberance of a peacock's tail would not exist without the peahen's ability to appreciate its magnificent size and shape and colour.

Of our five senses, sight is considered by most people the most valuable, and our ability to see the world in three dimensions and in full colour is surely one of the most extraordinary achievements of evolution. The complex design of a vertebrate eye has long been seen as a challenge to Darwin's theory of evolution; Darwin himself conceded that its intricacy might seem to be 'absurd in the highest possible degree'.[7] To make a vertebrate eye required the invention of, inter alia: a precisely shaped and perfectly transparent lens made of a huge crystal of protein molecules; surrounding muscles able to move and shape and focus the lens; an iris to regulate the quantity of incoming light; more than 100 million photoreceptor cells arrayed across the retina to capture the incoming photons; and, perhaps most astounding of all, a brain able to translate a series of electrical impulses arriving along the optic nerve into a comprehensible, full-colour, three-dimensional representation of the world about us.

The workings of the brain and our consciousness of vision are certainly the least understood of the stages of seeing, but, from the point of view of physics, just as extraordinary is our eye's ability to detect light at all – massless photons barrelling towards us at the speed of light. And even cleverer is our eyes' ability to distinguish between photons with subtly different wavelengths – red and blue and green – meaning that the movie of our lives plays in glorious Technicolor. The story of how animals detect photons is now known in some detail. The key that unlocked the mystery was the discovery that the proteins that detect light are a tiny twig on a much more expansive family tree that reveals the relationships between a great many similar proteins. But first a little background . . .

Knowing that genes occasionally jump from one branch to another tells us that we cannot always rely on genes to tell us about the relationships between species. To make matters worse, sometimes a gene becomes duplicated. From the moment of duplication, the two copies live as independent entities; while they start their independent lives as identical twins, the two are immediately free to accumulate their own mutations with no reference to the other, and both copies of a duplicated gene get inherited in parallel as the tree of life grows and new species form. Gene duplication has turned out to be very common; for mammals at least, roughly 200 of our 20,000-odd genes get duplicated every million years.[8]*

Familiar examples of duplicated genes have been seen already in the Hox genes, produced by repeated duplications of a 600-million-year-old founder gene. Likewise, all mammals have several copies of the globin genes, which code for: the oxygen-carrying haemoglobin protein found in all our adult red blood cells; a similar but different version of haemoglobin used only in the red blood cells of a foetus in the womb; and myoglobin, found in our muscles (where it stores oxygen).[9] All of these globin genes are obviously related, a fact that can easily be read in their similar functions and in the similar sequences of amino acids that they produce.

Why might we care about knowing the evolutionary history of genes that are duplicated in our genomes? The general answer is that gene duplication is a rich source of novelty – the 1,000 or so duplicated genes that humans have accumulated since we parted company from the chimpanzees must, for example, explain some of the differences between us. More emphatically, the unique complexity of vertebrates and, in particular, the invention of the

* This rate of duplication does not mean we end up with more and more genes because the rate of new genes appearing due to duplication is approximately matched by the reverse process whereby genes are lost. Duplicates, which, at least straight after the duplication, are identical to each other with identical functions, are naturally the most dispensable of all the genes in the genome and so, until they develop their own personalities and tasks, are the most likely to be lost.

vertebrate head have been linked, as we have seen, to the duplicating of the entire genome of our ancestor and not once, but twice. Early in our evolution, every single gene in the genome of the vertebrate ancestor suddenly became present in four copies, each one able to go in its own direction, providing new variations on the basic vertebrate body for natural selection to choose amongst.

To see up close a concrete example of what gene duplication can achieve, let's get back to the evolution of vision. In the retina, right at the back of your eyeball, there is a large and dense cluster of cells called photoreceptors, like a tightly planted field of sunflowers, whose sole job is to gather light. Photoreceptor cells are not unique to vertebrates and can be found in one form or another in all the many kinds of eye that exist across the animal kingdom. Each photoreceptor cell resembles a tall, narrow tower that is packed with a stack of tiny discs arranged so that their flat sides cut across the rays of light entering the eye. Embedded in these discs, in their tens of thousands, are proteins called opsins, which are responsible for capturing and detecting photons.

If you read the sequence of amino acids in the opsin protein and compare this to the sequences of all the other human genes, you will find a host of similar-looking genes. In fact the human genome contains about 800 members of this family of opsin-like genes.[10] There is really only one plausible explanation for the existence of this family of hundreds of similar genes, and this is that they all originated at some point in the deep past from a single ancestor gene. This prototype must have been duplicated over and over again throughout the evolutionary history of humans to produce more and more members of this family of proteins.

While only a handful of these 800 human genes – which are collectively named G protein-coupled receptors (GPCRs) – have anything to do with sensing light, it is nevertheless possible to grasp that their tasks are at some level similar. As their name reveals, GPCRs are receptors – which means that there is a part of the protein that sticks out of the cell like a satellite receiver on the side of a house. The function of the receiver is to capture and hence recognise the presence of one specific molecule, and this allows our cells to sense what is going on in the environment around them. The most familiar function of GPCRs is actually found in the cells of our taste buds. One kind of taste bud cell has a receiver into which a sugar molecule fits perfectly (detecting sweetness); a second is a fit for the amino acids in the protein that we eat (allowing us to taste savoury or 'umami' flavours); a third detects molecules associated with bitterness.*

Gathering all the amino acid sequences produced by all the GPCR genes that are to be found in our genome, we can, just as if they were species, find out how they are all related.[11] There are two important messages that come from this tree of GPCR genes. The first is that all the light-detecting animal opsins are grouped together as a single branch on the much larger GPCR tree. Opsins, therefore, with their ability to detect photons, evolved just once in a long-dead ancestor of the animals.

The second notable result is that of all the hundreds of types of non-light-detecting GPCRs, there is one that is most closely related to the light-sensing opsins. This is the receptor that detects melatonin, a name you probably recognise. Melatonin is a hormone secreted by a gland in our brains in response to the time of day; it is a signal sent out in our bloodstream to all the cells of our body that tells them that it is time for sleep or time to wake up. Our cells need to detect the melatonin circulating in our blood in order to behave appropriately (get up and brew the coffee or brush teeth and put on pyjamas), and they do so using a GPCR called, sensibly enough, the melatonin receptor.

The close relationship between opsin GPCRs and the melatonin receptor GPCR has turned out to be a clue telling us just how opsins became able to detect photons. The chain of events must have begun with a single parent gene (one that was very likely already detecting melatonin). Perhaps 600 million years ago in the ancestor of all animals, this parent gene duplicated to produce two daughter genes, and these two daughters were now free to

* Sourness and saltiness are detected by other kinds of protein.

follow separate paths; one continued to detect melatonin and the other changed in some fashion to develop the amazing new ability to detect light.

While opsins are used to sense light, what they actually detect is a specific form of a small molecule called retinol. Retinol, also known as vitamin A1, is similar to melatonin in size and chemical structure. The retinol molecule sits permanently in the receptor bit of the opsin protein. It is deemed to be inactive and, in this state, is effectively invisible to the opsin. But when an inactive retinol molecule is hit by a photon, its shape suddenly changes, and it is this second active form of the molecule that can be detected by the opsin. The alerted opsin now sends out the signal that a photon has landed, and this message, ultimately, ends up in the brain. In this ingenious way, a receptor protein was repurposed so that we (and all animals) could detect light.

The invention of a photoreceptor cell with its magical opsin proteins was like adding wheels to a Lego set that had only consisted of bricks. The different uses to which photoreceptors can be put are limited only by evolution's imagination. Their simplest use is simply to let an animal know whether it is light (daytime) or dark (night-time). A tiny planktonic animal in the sea might want to rise to the surface in daylight to soak up the sun and sink at night to avoid being eaten. Only slightly more sophisticated, a flatworm, seeking shelter under a stone, might want to detect not just the general presence of light but its direction so as to swim towards or away from it – this is a skill that can be achieved with just two photoreceptor cells, one pointing left and one pointing right, and by applying a simple rule: to find the dark, steer towards the eye that detects fewer photons. A further step up in sophistication results in an image – objects and shapes rather than just light or dark. This requires several photoreceptors working together. Like adding pixels to a smartphone to get a sharper picture, the more photoreceptors in your eyes, the clearer the image it perceives: a human retina has a healthy 200,000 photoreceptors per square millimetre, but an eagle squeezes in five times as many.

The most sophisticated eyes of all are not only acute like an eagle's but contain more than one kind of photoreceptor, each one making its own special opsin that reacts most strongly to photons of a certain wavelength or colour. Whales, with a single opsin, see the world in greyscale; most other mammals, with two opsins, detect shades of blues and greens; humans have a third, red-sensitive opsin to gain access to yellows and reds and oranges; amphibians, birds and reptiles have a fourth and can perceive colours in the 'ultraviolet' range that we cannot imagine. The royals amongst the vertebrates may be the odd couple of pigeons and lampreys, which seem to have five.[12] And the wonderful result has been the exploitation of these colour-sensitive eyes by other species across the tree of life. Eyes were the spur for the evolution of parrots' feathers; ripe red fruit; the warning stripes of wasps; and the flashing, pulsating skin of a cuttlefish.

This chapter has taken us to the very highest point of the tree of life to think about three parts of biology – our cells, our genes and our culture – that seem to rise beyond the topmost leaves. To end, I want to think about three much more intangible parts of the tree of life: the unseen branches from the past; the undiscovered branches that exist today; and, last of all, the unknowable branches of the future.

23

Unknown Unknowns and the End of the Affair

IT'S FEBRUARY 2024, and I am travelling with some colleagues (and my youngest daughter, Francesca) to a town called Río Cuarto, which sits on the endless flat farmlands of Córdoba province to the east of the Argentinian Andes. This is a small and unremarkable city, no one's idea of a tourist destination; we are here not for sight-seeing but to search for a worm that might just contain the potential to rewrite textbooks. In truth, is it more likely that our expedition is a wild goose chase – in search of a worm that existed only in the imagination of its supposed discoverer. The mundane middle ground would be to discover that this tiny organism exists but is not in fact an important worm, nor even an animal, but a mere cluster of ciliated cells linked together in a small, worm-shaped colony. This worm (let's be optimistic) has no common name – indeed it is arguably the least common of all described animals, having been seen by one person ever at most – but it has a scientific name: *Salinella salve.**

Salinella was (perhaps) discovered and described by the nineteenth-century German zoologist Johannes Frenzel in a series of papers published, in German and in English, in 1891 and 1892.[1] The dates are important; these years coincide with the height of Ernst Haeckel's fame and influence, one result of which was the spreading amongst zoologists of the passionate hope of

* *Salinella* means something like 'little salt dweller' and *salve* means 'hello', perhaps commemorating the first expression of startled welcome that escaped its discoverer's lips as it came into focus under his microscope.

discovering living survivors of the earliest stages of animal evolution. Frenzel was born in Prussian Posen (now Polish Poznań) in 1858.[2] He was bright and successful, studying at distinguished universities in Berlin and Göttingen, and considered by his professors to be a 'particularly diligent and active student'. He was appointed, on 1 April 1887, to a professorship at the University of Córdoba, where he hoped to focus his efforts on describing the little-known, microscopic, single-celled life of Argentina. A photograph of him aged thirty-seven shows a face whose only remarkable features (despite my best efforts to detect the malevolent sign of a man who might invent a worm) are his intense eyes; he appears otherwise as just another middle-class, Mitteleuropean professional with a beard and the first hint of an exciting moustache. By the time the first *Salinella* paper appeared in print, Frenzel had already left the intellectual backwater of Córdoba (apparently fed up with the obligation to work on beetles and birds and bemoaning the lack of books) and returned to Germany. Sadly, he did not last much longer; after founding and briefly leading a fisheries institute on Lake Müggelsee close to Berlin, he died in 1897 at the age of thirty-nine, by falling off a bridge.

If we take at face value Frenzel's description and interpretation of *Salinella*, it is the very simplest living animal of all. And its simplicity is key to its importance, understood by Frenzel as revealing *Salinella* to be our very most distant animal relative, isolated from humans and indeed from all other animals by perhaps 600 million years of evolution. Frenzel pictures *Salinella* as a vital missing link, 'the first and only example of a connecting-link between Protozoa and Metazoa'.[3] If this is true, it would be very exciting indeed – a portal capable of whisking us back to the moment in our very distant past when our ancestors were taking the very first steps of the animal experiment: a tiny simple body of cells that contained the idea of a butterfly and a whale.

Frenzel describes *Salinella* as a microscopic, slightly flattened cylinder made from a single sheet of cells that has been rolled up to form a tube. He identifies a mouth at one end and an anus at the other and little else beyond a tuft of cilia wafting around each of these openings – no eyes, no brain, no muscles, nerves, organs, backbone, testes, ovaries . . . barely an animal at all. To a very good approximation, *Salinella* is just a gut.

Frenzel's several papers describing *Salinella* are a strange mix of, on one hand, nineteenth-century German competence – careful description and scientific professionalism – and, on the other, an account of a chaotic discovery and the least reliable 'methods section' you will ever read in a scientific paper. The story of the discovery goes that a friend of Frenzel – Dr Wilhelm 'Guillermo' Bodenbender (aka 'Bodenbender the tireless')[4] – collected mixed salt and soil samples 'from the salt pans in the Río Cuarto region in the south of the province of Córdoba' (helpfully narrowing our own search down to a few hundred square kilometres I suppose).[5] Frenzel, looking, presumably, for the salt-loving, single-celled ciliates he was an expert on, dissolved this salty soil in tap water (20 g of salt per litre). The next step is actually pretty funny: apparently by accident (and possibly importantly or perhaps not) 'a very small quantity of a highly diluted iodine solution had also got into it' (a drop? a splash? an egg cup full? 'got into it'!). A more haphazard and less rigorous account of a scientific method is hard to imagine. The salty, iodiney tap water was then left on a windowsill, and the next part of the 'protocol' (a generous term) may be the hardest to replicate of all: the container was 'sometimes open, sometimes covered, semi-exposed and exposed to the sun for a short time each day'. It gets worse . . . 'Dust and sand, dead flies, etc., had likewise fallen in abundantly.'[6] To make replication even more impossible, Frenzel also added unspecified quantities of an unidentified pond weed, and the water was topped up from time to time over a period of three months as it evaporated. The actual source of the *Salinella* he found is entirely opaque – was it in the salt from the salt pans or in the soil? Where did these samples even come from? Did it blow in through the window in the dust and sand or on a dead fly? Or did he add it by chance along with the pond weed?

The suspicion that *Salinella* is a work of fiction – a convenient 'discovery' of a longed-for ancestral animal type – seems a reasonable place to start. Personally, I veer between scepticism and hope: on one side the multiple, long and detailed papers surely too elaborate for a hoax; on the other, the precious samples falling apart when iodine was added making them unexaminable by anyone else rather too convenient. But the appeal of rediscovering a lost and possibly pivotal animal phylum (and of studying its morphology and DNA to work out where it really belongs on the tree of life) was enough to spur us to make the trip to Córdoba province.

We were not the first to follow the clues in Frenzel's manuscripts to look for the lost worm of the salt pans. The German zoologist Professor Michael Schrödl had the same thought over twenty years ago. Schrödl came back disappointed (not least at the lack of any obvious salt pans anywhere near Río Cuarto). Our own search thus far has been, I am sorry to report, just as unfruitful; our trip perhaps most useful in showing us just how hard it is going to be to find the source of Bodenbender's sample in a landscape transformed in the past 130 years from boundless Pampas to equally endless fields of soya, alfalfa and maize.

Salinella remains, at least, a known unknown, a zoological version of other mythical, expedition-prompting South American treasures like El Dorado and the Sierra de la Plata. But even if *Salinella* is a figment of Frenzel's imagination, we can say with total confidence that there are today a great many branches of the tree of life whose existence we have no inkling of whatsoever – a version of Donald Rumsfeld's often mocked 'unknown unknowns – the ones we don't know we don't know'.[7] These modern-day mysteries are in their turn outnumbered many thousands of times over by a host of extinct forms of life. It feels slightly frustrating to realise that these unknown unknowns mean that we will never complete the tree of life, and this imperfect tree means that our time machine will be forever restricted in where it can travel. Thanks to the many gaps – which get bigger the further back in time we go – we will never be able to meet every ancestor.

Humans have so far described and named a little more than a million living leaves on the tree of life, but estimates of the true number of species begin at about 9 million (and go as crazily high as a trillion, you will recall).[8] Conservatively, then, at least 8 million species alive today are unknown to us – the great majority in fact. The existence of at least some of these shadows can be found in the traces they leave behind in the environment, like evidence at a crime scene, in the form of their DNA. If we read the sequences of nucleotides in the DNA that can be filtered out of any given litre of seawater, or extracted from a spoonful of mud from the bottom of any river, and compare them to those in our databases of known species (think Interpol's database of criminals), we find a huge number of DNA sequences we don't recognise, from species of bacteria, fungi, plants and animals we have never seen.

To grasp the breadth and depth of our ignorance with a minor but perhaps by now familiar branch of the tree, the team of the Spanish biologist Professor Iñaki Ruiz-Trillo collected the mix of genes coding for small subunit ribosomal RNAs that could be found in seawater sampled from various different marine environments. They looked amongst these for the ribosomal RNAs coming just from one small group of animals – the acoel worms (the mint sauce worm *Symsagittifera*, and its relatives). Ruiz-Trillo's team found ribosomal RNAs that came from over 100 different acoel-like animals; seventy-five of these belonged to completely unknown species.[9] Other bits of the tree of life are even less likely to be known from a species we have seen: from a specimen we have in a bottle of formaldehyde or fixed on a microscope slide. Almost all the evidence we have of the existence of the archaean asgards – the closest living relatives of the cells that formed the first eukaryotes – comes from environmental DNA and, out of the many asgardian species whose DNA we can detect, only a minuscule proportion have ever been seen in the flesh, still fewer grown in the lab.

Even at a million-odd species, then, the image of the diversity of life on earth that we have today is much closer to an initial sketch than to a finished picture. Some parts are undoubtedly almost complete – it's unlikely that there are hundreds of unknown mammals or lizards or birds to be discovered – but other parts are all but blank. These invisible species are the tiny, the slow-growing, the awkward to culture and the difficult to collect – a description, in fact, that fits *Salinella* well enough to give us a small hope that maybe, if we can only look hard enough, we will find it.

If the list of the invisible branches that exist today is long, the list of the lost must be thousands of times longer. Some extinct species are hugely abundant in the fossil record: ammonites, belemnites, bivalves, brachiopods and trilobites; shark teeth and whole Devonian fishes; sea urchins and the disarticulated, five-sided stalks of sea lilies to name check just some animals. But so many other groups of animals and plants are incredibly rare, tiny and soft and all but impossible to fossilise; some may have lived in environments unlikely to make fossils (mountaintops or deserts, for example); some species have tiny numbers of individuals; others were simply short-lived. Single-celled organisms are much, much less likely to have made a permanent imprint in the rocks than a shark or an ammonite. The exceptions being the big structures such as stromatolites that they can build when they club together or, even bigger, the huge chalk cliffs made from countless mineral-rich skeletons of tiny coccolithophores and foraminiferans.

Most inaccessible of all past life, however, are the entities that came before LUCA. This blank on our canvas is frustrating and exciting in equal measure: working out the events on the early earth that made life from rocks, biochemistry from chemistry, is one of the most interesting problems in all of science and one of the most difficult.

Life's appearance should never be thought of as a single event; LUCA, though simple, tiny and unicellular, with no clever thoughts, feelings or even behaviour to speak of, was nevertheless already a fearsomely complex organism, the product of a long gestation. LUCA must have had a beautifully optimised biochemistry that allowed it to thrive in the hostile conditions of the early earth and to begin its 4-billion-year dynasty. It had DNA and RNA, proteins, enzymes, cell membranes; it had, you may remember, a reverse gyrase protein to twist up its DNA and iron-sulphur proteins and ribosomes and a genetic code. All of these complex parts could not have magically appeared in one go, but the order of their appearance is opaque because we do not (cannot) know what came before LUCA. Today all these qualities of LUCA – inherited by all the life that sprung from it – are tightly integrated and interdependent, each one essential, making it difficult to imagine how they could have been assembled one by one.

For these very oldest events in the emergence of life, our two paths to studying the characteristics of past organisms – looking in the fossil record and extrapolating from living descendants – are both blocked. Few rocks and certainly no fossils have survived the aeons of erosion, plate tectonics and volcanism that followed the origins of life; and, with just a single lineage remaining of the many that must have existed prior to LUCA, our method of backward extrapolation hits a roadblock 4 billion years ago.

The fearsome difficulty of reconstructing the steps involved in the origins of life seem to have made this field the wild west of evolutionary (or even pre-evolutionary) biology – crazy ideas are a *sine qua non.* But we can place three constraints on any viable theory in order at least to narrow down the versions of events. The first is that the events proposed must have taken place in the conditions, as we understand them, of the early earth: the likely chemical composition of the ocean and the atmosphere, the temperature of these environments, the acidity, the salinity. All of these constrain the chemical reactions that could have occurred. For seventy years, experiments have tried to recreate plausible conditions of the young earth to see what chemical reactions are possible and what precursors of biological molecules might spontaneously assemble – amino acids, sugars, nucleotide bases. The second constraint comes from the need to provide the universal requirements of life. The theory needs to include a source of energy; for evolution to begin, a mechanism for inheritance of characteristics (not necessarily DNA) needs to exist; and there needs to be a way of enclosing the constituents of 'individuals' so everything doesn't just leak away (not necessarily a cell membrane). The final constraint is that, whatever the series of events may be, we must end up not with just any form of life, but specifically with LUCA.

Even if all past and present species were somehow discovered and described, our tree of life would remain incomplete. The tree of life is only part grown, much more than a sapling but still some way from maturity, and the fate of the branches that are alive today is pretty much unknowable. The evolution of new branches on the tree of life (and the extinction of others) is only predictable on very short timescales and in rather vague ways – we expect that big branches with many leaves and species that are jacks of all trades are likely to endure whatever the future throws at them; small groups of species and ultra-specialists are more likely to disappear. But, because the totality of earth's ecosystem is so complex and chaotic, more specific predictions, and especially those on a slightly longer timescale, are futile – no one could have anticipated that the coming together of an archaean and a eubacterium would ultimately produce mushrooms, giant redwoods and humans, or foreseen the blossoming of birds and mammals thanks to a chance encounter with an asteroid.

We can make one safe prediction, however: the only real certainty for the future of the tree of life on earth is that it will one day stop growing. The tree must wither and die when the planet inevitably becomes so inimical to life that nothing, not even the hardiest of extremophile bacteria, can survive on it. The end of the habitability of the planet is closer than you might think. Ignoring unpredictable alternatives such as a nearby supernova, asteroids or the disruption of the earth's orbit by a passing star, our life-giving sun is the most likely culprit for the end of the tree of life. In about 5 billion years, the hydrogen fuel in the sun's core will be exhausted; the slowing of the nuclear fusion at its centre will result in the collapse of the core and the beginning of a new set of fusion reactions in the gases outside the core. These reactions will cause the sun to swell to become a red giant star that is so massive it will entirely engulf the orbits of Mercury and Venus and, possibly, the earth.

Five billion years in the future doesn't sound too bad – we are only halfway there. But the tree of life will not survive to witness this event, life's end will come much sooner and from something that seems rather benign, like the slowly warming water that surrounds a lobster in a pot. The change comes from the brightening of the sun, just a 1% increase every 100 million years, as it fuses light hydrogen atoms into slightly heavier helium, resulting in a denser and hotter core.[10] Recent work predicts that when the intensity of sunlight has increased by about 12%, the climate of the planet will change quickly to become dangerously hot.

According to these models, as the planet gently warms, more of the waters of the oceans will be converted into vapour in the atmosphere, where it will act as a greenhouse gas. Eventually, a tipping point will be reached when the extra warmth coming from the greenhouse effect is itself the cause of more evaporation from the ocean, which in turn leads to even more greenhouse warming. Once this positive feedback loop is established, the planet will take a slow-motion leap (lasting a couple of hundred million years) from today's comfortable average global temperature of about 17 °C to a scorching 55 °C. Three billion years from now, all of earth's water will have boiled off into space, signalling a definitive end for almost all life on earth.

The bad news continues for the descendants of the human race (assuming we haven't conspired to kill ourselves in the meantime), who may have much less than a billion years left. For starters, there will be barely any oxygen in the atmosphere in a billion years' time. But things get nasty, for mammals at least, much sooner than this. Predictions of the future movement of the continents shows they are destined to drift towards each other to form, in just 250 million years, a new supercontinent that has been named Ultima Pangea.[11] The volcanic activity caused by these tectonic collisions will collude with the brightening sun to produce a continent too hot for any mammal to live on. Marine animals will presumably survive longer; rats will outlast pandas; cockroaches will outlast butterflies; and hardy bacteria and archaeans living in hydrothermal vents at the bottom of the oceans may be the most tenacious of all. But there will be a day – an instant even – on which the very last individual of the very last species of life on earth dies. In that moment, the definitive and complete tree of life on earth can be drawn. But, of course, there will be no one around to do so.

The inevitable death of the tree of life may seem a deeply sad idea, but an omnipotent and eternal being able to take stock of what has happened over perhaps 5.5 billion years on this ordinary little planet could only look back on it as the most extraordinary adventure. Such a being would have gazed down as a series of amazingly improbable events conspired to produce, from the chemicals and the energy in an undersea volcano, the very first living things – perhaps a unique event in the vastness of the universe. And then, this first domino having toppled, this god will have witnessed the release of the unstoppable genie of evolution, a process that allowed this first precarious seed not just to survive but to blossom in a most extraordinary way. The all-seeing god will have watched the branching tree grow, each new twig tantalising with the possibility of what marvels it might produce, what huge bough it might one day become, all the time holding their breath at the thought that this branch, the whole tree even, might at any moment wither and die. The product of this phenomenon has been, is, and will continue to be the amazing variety of the life that surrounds us. At last, the full-grown tree of life will describe the life and death of the most complex machines ever built, from simple cells to thoughtful human beings to vast, planet-changing ecosystems.

Acknowledgments

I HAVE HAD A lot of help writing this book. I will try and thank people in roughly chronological order. My first thanks (it may surprise him) go to Henry Gee, who, when I sent him an over-ambitious proposal for a *Nature* review on how to use a tree of life, suggested it might be more suitable as a book. Next, thanks go to Jenny McCartney who began the process of steering my idea for this book – at first a bit of a textbook – into something that people might enjoy reading. Great thanks are due to my agent Will Francis and colleagues at Janklow and Nesbit UK for taking me on, and most of all for providing encouragement when it was most needed. To my UK publishers John Murray, most especially to my editors Georgina Laycock and Kate Craigie who finished the job that Jenny began; to Caroline Westmore who kept me on the straight and narrow; and to Sam Wells for his fantastic copy-editing. Thanks also to my American editor Jessica Yao at W. W. Norton in New York for some essential suggestions.

The first draft of the book was written during a two-term visiting fellowship at All Souls College, Oxford. This was a unique opportunity to spend a lot of time writing, and the generosity and friendly welcome from the warden, Professor Sir John Vickers, and all of the fellows I spent time with will never be forgotten. The other really important part of my time in Oxford was the other visiting fellows – in alphabetical order: Sophie Ambler, Nancy van Deusen, Coulter George, John Keown, James Lee, Sara Lipton, Érico Nogueira, Drazen Prelec, Rubina Raja, Jennifer Richards, Navtej Sarna, Laura Schaposnik, Matthew Syed and John Wyver. Thank you all so much!

I have to thank a host of readers; first amongst these is my friend Richard Copley, who read the entire book and helped me avoid numerous errors. Also, many thanks for advice and comments goes to colleagues Julia Day, Seirian Sumner, Helen Robertson, Paschalia Kapli, Tomáš Flouri, Irepan Salvador-Martínez, Ana Serra-Silva and Duncan Greig. The faults remaining are my own of course. I need to extend a very warm general thanks to my wonderful colleagues at UCL and especially to the members of my lab past and present. Finally, to my family, Lorna, Celia, Seth and Francesca for tolerating my stress towards the end of this process, which was a lot harder than expected.

Sources and Bibliography

Chapter 1: Solving Science's Greatest Puzzle

1. Roberts, S. (2020). 'Darwin's missing notebooks'. Cambridge University Libraries, https://www.cam.ac.uk/stories/darwin-appeal
2. Archibald, J.D. (2014). *Aristotle's Ladder, Darwin's Tree: The Evolution of Visual Metaphors for Biological Order.* Columbia University Press, New York; Pietsch, T.W. (2012). *Trees of Life: A Visual History of Evolution.* Johns Hopkins University Press, Baltimore.
3. Darwin, C. (1859). *On the Origin of Species by Means of Natural Selection, or The Preservation of Favoured Races in the Struggle for Life.* John Murray, London. p. 129.
4. Hopwood, T. (1959). 'The development of pre-Linnaean taxonomy'. *Proceedings of the Linnean Society of London* 170:230–4.
5. Aristotle, *De Partibus.* Quoted in: Hopwood, T. (1959). 'The development of pre-linnaean taxonomy'. *Proceedings of the Linnean Society of London,* 170:230–4.
6. Bonnet, C. (1764). *Contemplation de la nature.* Marc Michel Rey, Amsterdam.
7. Hellström, N.P. (2012). 'Darwin and the Tree of Life: the roots of the evolutionary tree', *Archives of Natural History* 39.2:234–52.
8. Augier, A. (1801). *Essai d'une nouvelle classification des vegetaux.* Bruyset Ainé et Co., Lyon.
9. Darwin Correspondence Project, 'Letter no. 2143' (accessed 25 September 2024).
10. Kutschera, U. et al. (2019). 'Ernst Haeckel (1834–1919): the German Darwin and his impact on modern biology'. *Theory in Biosciences* 138:1–7.
11. Haeckel, E. (1866). *Generelle Morphologie der Organismen: allgemeine Grundzüge der organischen Formen-Wissenschaft, mechanisch begründet durch die von Charles Darwin reformirte Descendenz-Theorie.* Georg Reimer, Berlin.

12. Haeckel, E. (1868). *Natürliche Schöpfungsgeschichte*. Georg Reimer, Berlin.
13. Darwin, C. (1859). *On the Origin of Species*.

Chapter 2: The Venus Flytrap and Other Unlikely Relatives

1. Darwin Correspondence Project, 'Letter no. 2996' (accessed 25 September 2024).
2. Darwin, C. (1875). *Insectivorous Plants*. John Murray, London.
3. Ellis, J. (1768). 'A new sensitive plant discovered'. *The London Magazine, or, Gentleman's Monthly Intelligencer*. October 1768. pp. 522–4.
4. Hodge, F.W. (ed.) (1912). *Handbook of American Indians, North of Mexico*. Part II. Government Printing Office, Washington, DC.
5. Ellis, J. (1768). 'A new sensitive plant discovered'.
6. Letter from John Bartram to Peter Collinson, 29 August 1762. In: Berkeley, E. and Berkeley, D.S. (eds) (1992). *The Correspondence of John Bartram, 1734–1777*. University Press of Florida, Gainesville. pp. 569–70.
7. Letter from Peter Collinson to John Bartram, 30 June 1764. In: Berkeley, E. and Berkeley, D.S. (eds) (1992). *The Correspondence of John Bartram*. pp. 631–3.
8. Ellis, J. (1768). 'A new sensitive plant discovered'.
9. Gibson, T.C. and Waller, D.M. (2009). 'Evolving Darwin's "most wonderful" plant: ecological steps to a snap-trap'. *New Phytologist* 183:575–87.
10. Palfalvi, G. et al. (2020). 'Genomes of the Venus flytrap and close relatives unveil the roots of plant carnivory'. *Current Biology* 30:2312–20.

Chapter 3: A Distant Cousin from the Ocean's Depths

1. Harrison, T. (2011). *Paleontology and Geology of Laetoli: Human Evolution in Context: Fossil Hominins and the Associated Fauna*. Volume II. *Fossil Hominins and the Associated Fauna* (Vertebrate Paleobiology and Paleoanthropology). Springer Science & Business Media, Berlin.
2. Courtenay-Latimer, M. (1989). 'Reminiscences of the discovery of the coelacanth, *Latimeria chalumnae* Smith'. *Interdisciplinary Journal of the International Society of Cryptozoology* 8:1–11.

3. Johanson, Z. et al. (2006). 'Oldest coelacanth, from the Early Devonian of Australia'. *Biology Letters* 2:443–6.
4. Smith, J.L.B. (1939). 'A living fish of Mesozoic type'. *Nature* 143:455–6.

Chapter 4: The Real History of the Birds and the Bees

1. Jenner, E. (1824). 'Some observations on the Migration of Birds. By the late Edward Jenner M.D. F.R.S.; with an Introductory Letter to Sir Humphry Davy, Bart. Pres. R. S. By the Rev. G.C. Jenner'. *Philosophical Transactions of the Royal Society* 114.
2. Hedenström, A. et al. (2016). 'Annual 10-month aerial life phase in the Common Swift *Apus apus*'. *Current Biology*, 26:3066–70.
3. Chen, A. et al. (2019). 'Total-evidence framework reveals complex morphological evolution in Nightbirds (Strisores)'. *Diversity* 11:143.
4. Belon, P. (1555). *L'histoire de la nature des oyseaux, avec leurs descriptions, & naïfs portraicts retirez du naturel: escrite en sept livres*. Paris.
5. Owen, R. (1843). *Lectures on the Comparative Anatomy and Physiology of the Invertebrate Animals.* Longman, Brown, Green & Longmans, London.
6. Owen, R. (1860). 'Darwin on the Origin of Species'. *Edinburgh Review* 3:487–532.
7. Owen, R. (1857). 'On the characters, principles of division, and primary groups of the class Mammalia'. *Zoological Journal of the Linnean Society* 2:1–37.
8. Darwin Correspondence Project, 'Letter no. 2611' (accessed 25 September 2024).
9. Owen, R. (1843). *Lectures on the Comparative Anatomy and Physiology of the Invertebrate Animals.*
10. Owen, R. (1849). *On the Nature of Limbs.* John Van Voorst, London.

Chapter 5: Are We Still Fishes?

1. Cherry-Garrard, A. (1922). *The Worst Journey in the World.* Constable & Co., London.
2. Hennig, W. (1966). *Phylogenetic Systematics.* University of Illinois Press, Urbana.
3. Williams, D., Schmitt, M. and Wheeler, Q. (eds) (2016). *The Future of Phylogenetic Systematics: The Legacy of Willi Hennig.* Cambridge University Press, Cambridge.
4. Ibid.

5. Hennig, B. and Kluge, A. 'Willi Hennig'. Willi Hennig Society, http://cladistics.org/willi-hennig/
6. Hennig, W. (1975). '"Cladistic Analysis or Cladistic Classification"?: a reply to Ernst Mayr'. *Systematic Zoology* 24:244–56.

Chapter 6: Some Awfully Big Numbers

1. Chapman, A.D. (2009). 'Numbers of living species in Australia and the World' (2nd edn). Australian Biodiversity Information Services, Toowoomba, Australia.
2. Lewis, E.B. (1998). *Alfred Henry Sturtevant 1891–1970. A Biographical Memoir.* National Academies Press, Washington, DC.
3. Dobzhansky, T. and Sturtevant, A.H. (1938). 'Inversions in the chromosomes of *Drosophila pseudoobscura*'. *Genetics* 23:28–64.

Chapter 7: Such Stuff as Genes Are Made On

1. Crick, F.H.C. (1958). 'On protein synthesis'. *Symposia of the Society for Experimental Biology* 12:138–63.
2. Harris, J.I., Sanger, F. and Naughton, M.A. (1956). 'Species differences in Insulin'. *Archives of Biochemistry and Biophysics* 65:427–38.
3. Field, K.G. et al. (1988). 'Molecular phylogeny of the animal kingdom'. *Science* 239:748–53.

Chapter 8: Meet LUCA, the Last Universal Common Ancestor

1. Javaux, E.J. (2019). 'Challenges in evidencing the earliest traces of life'. *Nature* 572:451–60.
2. Bell, E.A. et al. (2015). 'Potentially biogenic carbon preserved in a 4.1 billion-year-old zircon'. *Proceedings of the National Academy of Sciences USA* 112:14518–21; Tashiro, T. et al. (2017). 'Early trace of life from 3.95 Ga sedimentary rocks in Labrador, Canada'. *Nature* 549:516–18.
3. Darwin, C. (1859). *On the Origin of Species.* p. 484.
4. Weiss, M.C. et al. (2018). 'The last universal common ancestor between ancient Earth chemistry and the onset of genetics'. *PLoS Genet* 14:e1007518.

5. Sapp, J. (2005). 'The prokaryote–eukaryote dichotomy: meanings and mythology'. *Microbiology and Molecular Biology Reviews* 69:292–305.
6. Luehrsen, K. (2014). 'Remembering Carl Woese', *RNA Biology* 11:217–19.
7. Kolter, R. (2024). 'Requiem for an apparatus'. Small Things Considered, https://schaechter.asmblog.org/schaechter/2024/03/requiem-for-an-apparatus.html
8. Pace, N.R. et al. (2012). 'Phylogeny and beyond: scientific, historical and conceptual significance of the first tree of life'. *Proceedings of the National Academy of Sciences USA* 109:1011–18.
9. Woese, C.R. and Fox, G.E. (1977). 'Phylogenetic structure of the prokaryotic domain: the primary kingdoms'. *Proceedings of the National Academy of Sciences USA* 74:5088–90.
10. Schwartz, R.M. and Dayhoff, M.O. (1978). 'Origins of prokaryotes, eukaryotes, mitochondria and chloroplasts'. *Science* 199:395–403.
11. Weiss, M. et al. (2016). 'The physiology and habitat of the last universal common ancestor'. *Nature Microbiology* 1:16116.
12. Moody, E.R.R. et al. (2024). 'The nature of the last universal common ancestor and its impact on the early Earth system'. *Nature Ecology and Evolution* 8:1654–66.
13. De Robertis, E. and Sasai, Y. (1996) 'A common plan for dorsoventral patterning in Bilateria'. *Nature* 380:37–40.

Chapter 9: Head-to-Tail Evolution and the First of the Animals

1. Lipshitz, H.D. (2005). 'From fruit flies to fallout: Ed Lewis and his science'. *Developmental Dynamics* 232:529–46.
2. Sturtevant, A.H. (1965, 2001). *A History of Genetics.* Cold Spring Harbor Laboratory Press, New York.
3. Morgan, T.H. (1910). 'Sex limited inheritance in *Drosophila*'. *Science* 32:120–2.
4. Lewis, E.B. (1978). 'A gene complex controlling segmentation in *Drosophila*'. *Nature* 276:565–70.
5. McGinnis, W. et al. (1984) 'A conserved DNA sequence in homoeotic genes of the *Drosophila* Antennapedia and bithorax complexes'. *Nature* 308:428–33.
6. McGinnis, W. et al. (1984) 'A homologous protein-coding sequence in *Drosophila* homeotic genes and its conservation in other metazoans'. *Cell* 37:403–8.

7. Holland, P.W.H. and Hogan, B.L.M. (1986). 'Phylogenetic distribution of Antennapedia-like homoeo boxes'. *Nature* 321:251–3.
8. Graham, A. et al. (1989). 'The murine and *Drosophila* homeobox gene complexes have common features of organization and expression'. *Cell* 57:367–78.
9. Small, K.M. and Potter, S.S. (1993). 'Homeotic transformations and limb defects in Hox A11 mutant mice'. *Genes Dev* 7:2318–28.
10. Gehring, W.J. and Ikeo, K. (1999). 'Pax 6: mastering eye morphogenesis and eye evolution'. *Trends in Genetics* 15:371–7.

Chapter 10: How the Insects Got Their Wings

1. Owen, R. (1843). *Lectures on the Comparative Anatomy and Physiology of the Invertebrate Animals.*
2. McMahon, D.P. et al. (2011). 'Strepsiptera'. *Current Biology* 21:R271–2.
3. Guillermo-Ferreira, G. and Gorb, S.N. (2021). 'Heat-distribution in the body and wings of the morpho dragonfly *Zenithoptera lanei* (Anisoptera: Libellulidae) and a possible mechanism of thermoregulation'. *Biological Journal of the Linnean Society* 133:179–86.
4. Boore, J. L. et al. (1998). 'Gene translocation links insects and crustaceans'. *Nature* 392:667–8.
5. Isabel Almudi, I. et al. (2020). 'Genomic adaptations to aquatic and aerial life in mayflies and the origin of insect wings'. *Nature Communications* 11:2631; Bruce, H.S. and Patel, N.H. (2020). 'Knockout of crustacean leg patterning genes suggests that insect wings and body walls evolved from ancient leg segments'. *Nature Ecology & Evolution* 4:1703–12.
6. Collins, D. (1996). 'The "evolution" of *Anomalocaris* and its classification in the arthropod class Dinocarida (nov.) and order Radiodonta (nov.)'. *Journal of Paleontology* 70:280–93.
7. Conway Morris, S. (1978). '*Laggania cambria* Walcott: a composite fossil'. *Journal of Paleontology* 52:126–31.
8. Nanglu, K. et al. (2022) 'Worms and gills, plates and spines: the evolutionary origins and incredible disparity of deuterostomes revealed by fossils, genes, and development'. *Biological Reviews* 98:316–51.
9. Whittington, H.B. and Briggs, D.E.G. (1982). 'A new conundrum from the Middle Cambrian Burgess Shale'. *Proceedings of the Third North American Paleontological Convention, Montreal* 2:573–5; Whittington,

H.B. and Briggs, D.E.G. (1985). 'The largest Cambrian animal, *Anomalocaris*, Burgess Shale, British Columbia'. *Philosophical Transactions of the Royal Society of London B* 309:569–609.

10. Van Roy, P. et al. (2015). 'Anomalocaridid trunk limb homology revealed by a giant filter-feeder with paired flaps'. *Nature* 522:77–81.
11. Usami, Y. (2006). 'Theoretical study on the body form and swimming pattern of *Anomalocaris* based on hydrodynamic simulation'. *Journal of Theoretical Biology* 238:11–17.
12. Daley, A.C. et al. (2009). 'The Burgess Shale anomalocaridid *Hurdia* and its significance for early euarthropod evolution'. *Science* 323:1597–600.

Chapter 11: The Microscopic Hitch-hikers Responsible for You and Me

1. Kostenko, A.G. and Mamkaev, Y.V. (1990). 'The position of "green convoluts" in the system of acoel turbellarians (Turbellaria, Acoela). 1. *Simsagittifera* gen. n. 2. Sagittiferidae fam. n.'. *Zoologicheskii Zhurnal* 69:11–21.
2. von Graff, L. (1891). *Die Organisation der Turbellaria Acoela.* Wilhelm Engelman, Leipzig.
3. Bailly, X. et al. (2014). 'The chimerical and multifaceted marine acoel *Symsagittifera roscoffensis*: from photosymbiosis to brain regeneration'. *Frontiers in Microbiology* 5:498.
4. Smith, J.A. and Ross, W.D. (eds) (1910). *The Works of Aristotle.* Volume IV: *Historia Animalium*, translated by D'Arcy Wentworth Thompson. Clarendon Press, Oxford.
5. Sagan, L. (1967). 'On the origin of mitosing cells'. *Journal of Theoretical Biology* 14:225–74.
6. Lake, J.A. (2011). 'Lynn Margulis (1938–2011), biologist who revolutionized our view of early cell evolution'. *Nature* 480:458.
7. Pizzorno, J. (2014). 'Mitochondria – fundamental to life and health'. *Integrative Medicine* 13:8–15.
8. Lazcano, A. and Peretó, J. (2017). 'On the origin of mitosing cells: A historical appraisal of Lynn Margulis endosymbiotic theory'. *Journal of Theoretical Biology* 434:80–7.
9. Sapp, J. et al. (2002). 'Symbiogenesis: the hidden face of Constantin Merezhkowsky'. *History and Philosophy of the Life Sciences* 24:413–40.

10. Merezhkowsky, C. (1903). *Das irdische Paradies. Ein Märchen aus dem 27 Jahrhundert: Eine Utopie*, translated by Helene Mordaunt. Friedich Gotthein, Berlin.
11. Schwendener, S. (1867). 'Uber die wahre Natur der Flechten'. *Verhandlungen der Schweizerischen Naturforschenden Gesellschaft in Rheinfelden* 5:88–90.
12. Virchow, R. (1858). *Die Cellularpathologie in ihrer Begründung auf physiologische und pathologische Gewebelehre.* August Hirschwald, Berlin.
13. Wallin, I.E. (1925). 'On the nature of mitochondria. IX Demonstration of the bacterial nature of mitochondria'. *American Journal of Anatomy* 36:131–49.
14. Lake, J.A. (2011). 'Lynn Margulis (1938–2011)'.
15. Sagan, L. (1967). 'On the origin of mitosing cells'.
16. Gray, M.W. and Doolittle, W.F. (1982). 'Has the endosymbiont hypothesis been proven?'. *Microbiological Reviews* 46:1–42.
17. Spencer, D.F. et al. (1984). 'Pronounced structural similarities between the small subunit ribosomal RNA genes of wheat mitochondria and *Escherichia coli*'. *Proceedings of the National Academy of Sciences USA* 81:493–7.
18. Sánchez-Baracaldo, P. et al. (2017). 'Early photosynthetic eukaryotes inhabited low-salinity habitats'. *Proceedings of the National Academy of Science USA* 114:E7737–45.
19. Williamson, D.I. (2009). 'Caterpillars evolved from onychophorans by hybridogenesis'. *Proceedings of the National Academy of Sciences USA* 106:19901–5.
20. Hart, M.W. and Grosberg, R.K. (2009). 'Caterpillars did not evolve from onychophorans by hybridogenesis'. *Proceedings of the National Academy of Sciences USA* 106:19906–9.
21. Plutarch (1914). *Plutarch's Lives*. Volume VII. William Heinemann, London.
22. Williams, T.A. et al. (2017). 'Integrative modeling of gene and genome evolution roots the archaeal tree of life'. *Proceedings of the National Academy of Sciences USA* 114:E4602–11.

Chapter 12: When Trees Go Wrong

1. Cucchi, T. et al. (2017). 'Detecting taxonomic and phylogenetic signals in equid cheek teeth: towards new palaeontological and archaeological proxies'. *Royal Society Open Science* 4160997.

2. Grant, P.R. and Grant, B.R. (2007). *How and Why Species Multiply: The Radiation of Darwin's Finches*. Princeton University Press, Princeton.
3. Seehausen, O. (2006). 'African cichlid fish: a model system in adaptive radiation research'. *Proceedings of the Royal Society B* 273:1987–98.
4. Jónsson, H. et al. (2014). 'Speciation with gene flow in equids despite extensive chromosomal plasticity'. *Proceedings of the National Academy of Sciences USA* 111: 18655–60.
5. Zink, R.M. and Vázquez-Miranda, H. (2019). 'Species limits and phylogenomic relationships of Darwin's finches remain unresolved: potential consequences of a volatile ecological setting'. *Systematic Biology* 68:347–57.
6. Lankester, E.R. (1885). 'The spiritualistic challenge'. Letter to *Pall Mall Gazette*, 13 January 1885. Quoted in: Milner, R. (1999). 'Huxley's bulldog: the battles of E. Ray Lankester (1846–1929)'. *The Anatomical Record (New Anat.)* 257:90–5.
7. Dawson, C. and Woodward, A.S. (1913). 'On the discovery of a palæolithic human skull and mandible in a flint-bearing gravel overlying the Wealden (Hastings Beds) at Piltdown, Fletching (Sussex)'. *Quarterly Journal of the Geological Society* 69:117–23.
8. White, L.C. et al. (2018). 'High-quality fossil dates support a synchronous, Late Holocene extinction of devils and thylacines in mainland Australia'. *Biology Letters* 14:20170642.
9. Haeckel, E. (1866). *Generelle Morphologie der Organismen*.
10. Kowalevsky, A.O. (1867). 'Entwicklungsgeschichte der einfachen ascidien'. *Mémoires de l'Academie des Sciences de St Pétersburg* 10:1–19.
11. Haeckel, E. (1868). *Natürliche Schöpfungsgeschichte*.
12. Giard, A.M. (1877). 'Sur les Orthonectida, classe nouvelle d'animaux parasites des Échinodermes et des Turbellariés'. *Comptes Rendus* 85:812–14.
13. von Kölliker, A. (1849). 'Üeber *Dicyema paradoxum*, den schmarotzer der venenanhänge der cephalopoden'. *Bericht der Königlichen Zootomischen Anstalt in Würzburg* 2:59–66.
14. Giard, A. (1880). 'The Orthonectida, a new class of the phylum of the worms'. *Quarterly Journal of Microscopical Science* (now *Journal of Cell Science*) 20(2):225–40.
15. Kristan Jr, W.B. et al. (2005). 'Neuronal control of leech behavior'. *Progress in Neurobiology* 76:279–327.

Chapter 13: Problems with Genes

1. Chen, Q. and Wang, Z. (2022). 'A new molecular mechanism supports that blue-greenish egg color evolved independently across chicken breeds'. *Poultry Science* 101:102223.
2. Whiting, M.F. and Wheeler, W.C. (1994). 'Insect homeotic transformation'. *Nature* 368:696.
3. Goldschmidt, R.B. (1940). *The Material Basis of Evolution*. Yale University Press, New Haven, Connecticut.
4. Specter, M. (1991). 'Hurricane rakes New England, loses some force'. *Washington Post*. 20 August 1991.
5. Huelsenbeck, J.P. and Hillis, D.M. (1993). 'Success of phylogenetic methods in the four-taxon case'. *Systematic Biology* 42:247–64.
6. Felsenstein, J. (1978). 'Cases in which parsimony or compatibility methods will be positively misleading'. *Systematic Zoology* 27:401–10.
7. Harcourt, A.H. et al. (1981). 'Testis weight, body weight and breeding system in primates'. *Nature* 293:55–7.
8. Marler, P. (1969). '*Colobus guereza*: territoriality and group composition'. *Science* 163:93–5.
9. Sugiyama, Y. (1971). 'Characteristics of the social life of Bonnet Macaques (*Macaca radiata*)'. *Primates* 12:247–66.
10. Markowitz, T.M. et al. (2023). 'Sociosexual behavior of nocturnally foraging Dusky and Spinner Dolphins'. In: Würsig, B. and Orbach, D.N. (eds). *Sex in Cetaceans: Morphology, Behavior, and the Evolution of Sexual Strategies*. Springer, Cham. Chapter 14.
11. Schwartz, J.H. and Tattersall, I. (2000). 'The human chin revisited: what is it and who has it?'. *Journal of Human Evolution* 38:367–409.

Chapter 14: Rock of Ages (or Ages of Rocks)

1. Geyer, G. and Landong, E. (2016). 'The Precambrian–Phanerozoic and Ediacaran–Cambrian boundaries: a historical approach to a dilemma'. In: Brasier, A.T., McIlroy, D. and McLoughlin, N. (eds). *Earth System Evolution and Early Life: A Celebration of the Work of Martin Brasier*. Geological Society, London, Special Publications. p. 448.

2. Rudkin, D.M. et al. (2003). 'The world's biggest trilobite – *Isotelus rex* new species from the Upper Ordovician of northern Manitoba, Canada'. *Journal of Paleontology* 77:99–112.
3. Morzadec, P. (2001). 'Les trilobites Asteropyginae de Devonian de l'Anti-Atlas (Maroc)'. *Palaeontographica Abteilung A*. 262:53–85.
4. Alvarez, L.W. et al. (1980). 'Extraterrestrial cause for the Cretaceous–Tertiary extinction'. *Science*. 208:1095–108.
5. During, M.A.D. et al. (2022). 'The Mesozoic terminated in boreal spring'. *Nature* 603:91–4.
6. Usserio, J. (1650). *Annales Veteris et Novi Testamenti a prima mundi origine deducti*. J. Flesher, London.
7. Beutner, D. (2022). 'Scientist and saint: blessed Niels Stensen (1638–1686)'. Catholic World Report, https://www.catholicworldreport.com/2022/12/05/scientist-and-saint-blessed-niels-stensen/
8. Hooke, R. (1705). *The Posthumous Works of Robert Hooke, Containing His Cutlerian Lectures, and Other Discourses, Read at the Meetings of the Illustrious Royal Society*. Richard Waller, London.
9. Poirier, J.-P. (2017). 'About the age of the Earth'. *Comptes Rendus Geoscience* 349:223–5.
10. Le Clerc, G.-L. (Comte de Buffon) (1774). *Histoire naturelle, générale et particulière*. Supplementary Volume 1. Royal Press, Paris. p. 158.
11. Kelvin, W.T. (Lord) (1864). 'On the secular cooling of the earth'. *Transactions of the Royal Society of Edinburgh* 23:167–9.
12. Richter, F.M. (1986). 'Kelvin and the age of the Earth'. *The Journal of Geology* 94:395–401.
13. Perry, J. (1895). 'On the age of the Earth'. *Nature* 51:224–7.
14. Davis, D.W. et al. (2003). 'Historical development of zircon geochronology'. *Reviews in Mineralogy and Geochemistry* 53:145–81.

Chapter 15: Using Our Genes to Tell the Time

1. Schleicher, A. (1853). 'Die ersten Spaltungen des indogermanischen Urvolkes'. *Allgemeine Monatsschrift für Wissenschaft und Literatur* 3:786–7.
2. Gontier, N. (2011). 'Depicting the tree of life: the philosophical and historical roots of evolutionary tree diagrams'. *Evolution: Education and Outreach* 5:15–38.
3. Newman, S. (1967). 'Morris Swadesh'. *Language* 43:948–57.

4. Swadesh, M. (1952). 'Lexico-statistic dating of prehistoric ethnic contacts: with special reference to North American indians and eskimos'. *Proceedings of the American Philosophical Society* 96:452–63.
5. Zuckerkandl, E. and Pauling, L. (1962). 'Molecular disease, evolution and genic heterogeneity'. In: Kasha, M. and Pullman, B. (eds). *Horizons in Biochemistry: Albert Szent-Gyorgyi Dedicatory Volume*. New York, Academic Press. pp. 189–225.
6. Mukharji, I. (2020). 'Emanuel Margoliash 1920–2008'. *Biographical Memoirs*. National Academy of Sciences, Washington, DC.
7. Margoliash, E. (1963). 'Primary structure and evolution of Cytochrome C'. *Proceedings of the National Academy of Sciences USA* 50:672–9.
8. Yochelson, E.L. (1996). 'Discovery, collection, and description of the Middle Cambrian Burgess Shale biota by Charles Doolittle Walcott'. *Proceedings of the American Philosophical Society* 140:469–545.
9. Wray, G.A. et al. (1996). 'Molecular evidence for deep Precambrian divergences among animal phyla'. *Science* 274:568–73.
10. Sperling, E.A. and Stockey, R.G. (2018). 'The temporal and environmental context of early animal evolution: considering all the ingredients of an "explosion"'. *Integrative and Comparative Biology* 58:605–22; Sun, W. et al. (2024). 'Developmental biology of *Spiralicellula* and the Ediacaran origin of crown metazoans'. *Proceedings of the Royal Society B* 291:20240101.

Chapter 16: Embryos and Arrow Worms

1. Uschmann, G. (1966). *Grobben, Karl: Neue Deutsche Biographie*. Historische Kommission bei der Bayerischen Akademie der Wissenschaften, Munich.
2. Toegel, C. (2024). *Sigmund Freud, 1856–1939: A Biographical Compendium*. Routledge, Abingdon.
3. Grobben, K. (1908). 'Die systematische Einteilung des Tierreichs'. *Verhandlungen der Kaiserlich-Königlichen Zoologisch-Botanischen Gesellschaft in Wien* 58:491–511.
4. Telford, M.J. and Holland, P.W.H. (1993). 'The phylogenetic affinities of the chaetognaths: a molecular analysis'. *Molecular Biology and Evolution* 10:660–76.
5. Marlétaz, F. et al. (2019). 'A new spiralian phylogeny places the enigmatic arrow worms among gnathiferans'. *Current Biology* 29:312–18.

Chapter 17: The Diet of Worms

1. Westblad, E. (1949). '*Xenoturbella bocki* n.g., n.sp., a peculiar, primitive Turbellarian type'. *Arkiv för Zoologi* 1:3–29.
2. Norén, M. and Jondelius, U. (1997). '*Xenoturbella*'s molluscan relatives'. *Nature* 390:31–2; Israelsson, O. (1997). 'and molluscan embryogenesis'. *Nature* 390:32.
3. Israelsson, O. (1999). 'New light on the enigmatic *Xenoturbella* (phylum uncertain): ontogeny and phylogeny'. *Proceedings of the Royal Society of London B* 266:835–41.
4. Telford, M.J. et al. (2024). 'Claus Nielsen (1938–2024), zoologist of invertebrates'. *Nature* 627:265.
5. Bourlat, S.J. et al. (2003). '*Xenoturbella* is a deuterostome that eats molluscs'. *Nature* 424:925–8.

Chapter 18: The First Three Billion Years

1. Mora, C. et al. (2011). 'How many species are there on Earth and in the Ocean?', *PLoS Biology* 2011;9:e1001127.
2. Locey, K.J. and Lennon, J.T. (2016). 'Scaling laws predict global microbial diversity'. *Proceedings of the National Academy of Sciences USA* 113:5970–5.
3. Holland, H.D. (2002). 'Volcanic gases, black smokers, and the great oxidation event'. *Geochimica et Cosmochimica Acta* 66:3811–26.
4. Hodgskiss, M.S.W. et al. (2019). 'A productivity collapse to end Earth's Great Oxidation'. *Proceedings of the National Academy of Sciences USA* 116:17207–12.
5. Probst, A.J. et al. (2014). 'Coupling genetic and chemical microbiome profiling reveals heterogeneity of archaeome and bacteriome in subsurface biofilms that are dominated by the same archaeal species'. *PLoS One* 27:e99801.
6. Davín, A.A. et al. (2023). 'An evolutionary timescale for Bacteria calibrated using the Great Oxidation Event'. *bioRxiv* 08.08.552427.
7. Woese, C.R. (1987). 'Bacterial evolution'. *Microbiological Reviews* 51:221–71.
8. Stackebrandt, E. et al. (1988). 'Proteobacteria classis nov., a name for the phylogenetic taxon that includes the "purple bacteria and their relatives"'. *International Journal of Systematic Bacteriology* 38:321–5.

9. Pedersen, R.B. et al. (2010). 'Discovery of a black smoker vent field and vent fauna at the Arctic Mid-Ocean Ridge'. *Nature Communications* 1:126.
10. University of Washington (2008). 'Scientists break record by finding northernmost hydrothermal vent field'. ScienceDaily, www.sciencedaily.com/releases/2008/07/080724153941.htm
11. Spang, A. et al. (2015). 'Complex archaea that bridge the gap between prokaryotes and eukaryotes'. *Nature* 521:173–9.
12. Woese, C. et al. (1990). 'Towards a natural system of organisms: proposal for the domains Archaea, Bacteria, and Eucarya'. *Proceedings of the National Academy of Sciences USA* 87:4576–9.
13. Williams, T.A. et al. (2017). 'Integrative modeling of gene and genome evolution'.
14. Eme, L. et al. (2023). 'Inference and reconstruction of the heimdallarchaeial ancestry of eukaryotes'. *Nature* 618:992–9.
15. Strassert, J.F.H. et al. (2021). 'A molecular timescale for eukaryote evolution with implications for the origin of red algal-derived plastids'. *Nature Communications* 12:1879.
16. Leadbetter, B.S.C. (2015). *The Choanoflagellates: Evolution, Biology and Ecology*. Cambridge University Press, Cambridge.
17. Vischer, W. (1945). 'Über einen pilzähnlichen, autotrophen Mikroorganismus, Chlorochytridion, einige neue Protococcales und die systematische Bedeutung der Chloroplasten'. *Verhandlungen der Naturforschenden Gesellschaft in Basel* 56:41–59.
18. Saville Kent, W. (1880). *Manual of the Infusoria: Including a Description of All Known Flagellate, Ciliate, and Tentaculiferous Protozoa, British and Foreign, and an Account of the Organization and Affinities of the Sponges*. David Bogue, London.

Chapter 19: The First Animals

1. Hooke, R. (1665). *Micrographia: or Some Physiological Descriptions of Minute Bodies Made by Magnifying Glasses. With Observations and Inquiries Thereupon*. J. Martyn and J. Allestry, Printers to the Royal Society, London.
2. Olivetta, M. et al. (2024). 'A multicellular developmental program in a close animal relative'. *Nature* 635:382–9.
3. Fairclough, S.R. et al. (2010). 'Multicellular development in a choanoflagellate'. *Current Biology* 20:R875–6.

4. Jékely, G. (2019). 'Evolution: how not to become an animal'. *Current Biology* 29:R1240–2.
5. Koehl, M.A.R. (2021). 'Selective factors in the evolution of multicellularity in choanoflagellates'. *Journal of Experimental Zoology (Molecular Development and Evolution)* 336:315–26.
6. Brunet, T. and King, N. (2017). 'The origin of animal multicellularity and cell differentiation'. *Developmental Cell* 43:124–40.
7. Richter, D.J. et al. (2018). 'Gene family innovation, conservation and loss on the animal stem lineage'. *eLife* 7:e34226; Coyle, M.C. and King, N. (2024). 'The evolutionary foundations of animal transcriptional regulatory mechanisms'. *Preprints* doi:10.20944/preprints202402.1653.v1
8. Koehl, M.A.R. (2021). 'Selective factors in the evolution of multicellularity in choanoflagellates'.
9. McIlroy, D. et al. (2024). 'The palaeobiology of two crown group cnidarians: *Haootia quadriformis* and *Mamsetia manunis* gen. et sp. nov. from the Ediacaran of Newfoundland, Canada'. *Life* 14:1096.
10. Telford, M.J. (2016). 'Fighting over a comb'. *Nature* 529:286.
11. Syed, T. and Schierwater, B. (2002). '*Trichoplax adhaerens*: discovered as a missing link forgotten as a hydrozoan, re-discovered as a key to metazoan evolution'. *Vie et Milieu / Life & Environment* 52:177–87.
12. Schulze, F.E. (1883). '*Trichoplax adhaerens* nov. gen. nov. spec.'. *Zoologischer Anzeiger* 6:92–7.
13. Sebé-Pedrós, A. et al. (2018). 'Early metazoan cell type diversity and the evolution of multicellular gene regulation'. *Nature Ecology and Evolution* 2:1176–88.
14. Varoqueaux, F. et al. (2018). 'High cell diversity and complex peptidergic signaling underlie placozoan behavior'. *Current Biology* 28:3495–501.e2.
15. Najle, S.R. et al., (2023). 'Stepwise emergence of the neuronal gene expression program in early animal evolution'. *Cell* 186:1–18.
16. Kapli, P. et al. (2021). 'Lack of support for Deuterostomia prompts reinterpretation of the first Bilateria'. *Science Advances* 7:eabe2741.
17. Grobben, K. (1908). 'Die systematische einteilung des tierreichs'. *Verhandlungen der Kaiserlich-Königlichen Zoologisch-Botanischen Gesellschaft in Wien* 58:491–511.

Chapter 20: The Road to Mammals

1. Conway Morris, S. and Caron, J.B. (2012). '*Pikaia gracilens* Walcott, a stem-group chordate from the Middle Cambrian of British Columbia'. *Biological Reviews* 87:480–512.
2. Walcott, C.D. (1911). 'Middle Cambrian annelids'. *Smithsonian Miscellaneous Collection* 57:109–44.
3. Mussini, G. et al. (2024). 'A new interpretation of *Pikaia* reveals the origins of the chordate body plan'. *Current Biology* 34:1–10.
4. Martik, M.L. and Bronner, M.E. (2021). 'Riding the crest to get a head: neural crest evolution in vertebrates'. *Nature Reviews Neuroscience* 22:616–26.
5. Graham, A. and Richardson, J. (2012). 'Developmental and evolutionary origins of the pharyngeal apparatus'. *EvoDevo* 3:24.
6. Brazeau, M.D. and Friedman, M. (2015). 'The origin and early phylogenetic history of jawed vertebrates'. *Nature* 520:490–7.
7. Nakatani, Y. et al. (2021). 'Reconstruction of proto-vertebrate, proto-cyclostome and proto-gnathostome genomes provides new insights into early vertebrate evolution'. *Nature Communications* 12:4489.
8. Narkiewicz, K. and Narkiewicz, M. (2015). 'The age of the oldest tetrapod tracks from Zachełmie, Poland'. *Lethaia Focus* 48:10–12.
9. Daeschler, E.B. et al. (2006). 'A Devonian tetrapod-like fish and the evolution of the tetrapod body plan'. *Nature* 440:757–63.
10. Clack, J.A. (2012). *Gaining Ground: The Origin and Evolution of Tetrapods*. 2nd edn. Indiana University Press, Bloomington, Indiana.
11. Coates, M.I. and Clack, J.A. (1990). 'Polydactyly in the earliest known tetrapod limbs'. *Nature* 347:66–9.
12. Coates, M.I. and Clack, J.A. (1991). 'Fish-like gills and breathing in the earliest known tetrapod'. *Nature* 352:234–6.
13. Clack, J. et al. (2017). 'Phylogenetic and environmental context of a Tournaisian tetrapod fauna'. *Nature Ecology and Evolution* 1:0002.

Chapter 21: The End of the Journey

1. Pickrell, J. (2019). 'The making of mammals'. *Nature* 574:468–72.
2. Mann, A. et al. (2020). 'Reassessment of historic "microsaurs" from Joggins, Nova Scotia, reveals hidden diversity in the earliest amniote ecosystem'. *Papers in Palaeontology* 6:605–25.

3. Zhou, C.F. et al. (2013). 'A Jurassic mammaliaform and the earliest mammalian evolutionary adaptations'. *Nature* 500:163–7.
4. Mao, F. et al. (2024). 'Fossils document evolutionary changes of jaw joint to mammalian middle ear'. *Nature* 628:576–81.
5. Hu, Y. et al. (2015). 'Large Mesozoic mammals fed on young dinosaurs'. *Nature* 433:149–52.
6. Beck, R.M.D. and Baillie, C. (2018). 'Improvements in the fossil record may largely resolve current conflicts between morphological and molecular estimates of mammal phylogeny'. *Proceedings of the Royal Society B* 285:20181632.
7. Meredith, R.W. et al. (2011). 'Impacts of the Cretaceous Terrestrial Revolution and KPg extinction on mammal diversification'. *Science* 334:521–4.
8. Korzh, V. and Grunwald, D. (2001). 'Nadine Dobrovolskaïa-Zavadskaïa and the dawn of developmental genetics'. *BioEssays* 23:365–71.
9. Xia, B. et al. (2024). 'On the genetic basis of tail-loss evolution in humans and apes'. *Nature* 626:1042–8.
10. Geissmann, T. and Bleisch, W. (2020). '*Nomascus hainanus*'. IUCN Red List of Threatened Species, e.T41643A17969392.
11. Brunet, M. et al. (2002). 'A new hominid from the Upper Miocene of Chad, Central Africa'. *Nature* 418:145–51.
12. Daver, G. et al. (2022). 'Postcranial evidence of late Miocene hominin bipedalism in Chad'. *Nature* 609:94–100.
13. Almécija, S. et al. (2021). 'Fossil apes and human evolution'. *Science* 372:587–99.
14. Vignaud, P., et al. (2002). 'Geology and palaeontology of the Upper Miocene Toros-Menalla hominid locality, Chad'. *Nature* 418:152–5.
15. Humphrey, L. and Stringer, C. (2018). *Our Human Story*. Natural History Museum, London.
16. Stringer C. (2016). 'The origin and evolution of *Homo sapiens*'. *Philosophical Transactions of the Royal Society B* 371:20150237.
17. Bergström, A. et al. (2021). 'Origins of modern human ancestry'. *Nature* 590:229–37.
18. Humphrey, L. and Stringer, C. (2018). *Our Human Story*.
19. Hu, W. et al. (2023). 'Genomic inference of a severe human bottleneck during the Early to Middle Pleistocene transition'. *Science* 381:979–84.
20. Greenspoon, L. et al. (2023). 'The global biomass of wild mammals'. *Proceedings of the National Academy of Sciences USA* 120:e2204892120.

Chapter 22: The Trees Within Us

1. Dawkins, R. (1976). *The Selfish Gene*. Oxford University Press, Oxford
2. Darwin, C. (1871). *The Descent of Man and Selection in Relation to Sex*. John Murray, London.
3. Pagel, M. (2017). 'Darwinian perspectives on the evolution of human languages'. *Psychonomic Bulletin & Review* 24:151–7.
4. Westcott, B.F. and Hort, F.J.A. (1881). *The New Testament in the Original Greek*. Harper & Brothers, New York.
5. Sulston, J.E. et al. (1983). 'The embryonic cell lineage of the nematode *Caenorhabditis elegans*'. *Developmental Biology* 100:64–119.
6. Félix, M.-A. and Sternberg, P.W. (1996). 'Symmetry breakage in the development of one-armed gonads in nematodes'. *Development* 122:2129–42.
7. Darwin, C. (1859). *On the Origin of Species*. Chapter 6.
8. Demuth, J.P. et al. (2006). 'The evolution of mammalian gene families'. *PLoS ONE* 1:e85.
9. Hardison, R.C. (2012). 'Evolution of hemoglobin and its genes'. *Cold Spring Harbour Perspectives in Medicine* 2:a011627.
10. Congreve, M. et al. (2020). 'Impact of GPCR structures on drug discovery'. *Cell* 181:81–91.
11. Feuda, R. et al. (2012). 'Metazoan opsin evolution reveals a simple route to animal vision'. *Proceedings of the National Academy of Sciences USA* 109:18868–72.
12. Emmerton, J. and Delhis, J.D. (1980). 'Wavelength discrimination in the "visible" and ultraviolet spectrum by pigeons'. *Journal of Comparative Physiology A* 141:47–52; Davies, W.L. et al. (2007). 'Functional characterization, tuning, and regulation of visual pigment gene expression in an anadromous lamprey'. *The FASEB Journal* 21:2713–24.

Chapter 23: Unknown Unknowns and the End of the Affair

1. Frenzel, J. (1891). 'Untersuchungen über die mikroskopische Fauna Argentiniens. Ein vielzelliges, infusorienartiges Their'. *Zoologischer Anzeiger* 367:230–3; Frenzel, J. (1892). 'Untersuchungen über die mikroskopische Fauna Argentiniens. *Salinella salve* nov. gen. nov. spec. Ein vielzelliges, infusorienartiges Tier (Mesozoon)'. *Archiv für*

Naturgeschichte 58(1):66–97; Frenzel, J. (1892). 'The mesozoon *Salinella*'. *The Annals and magazine of natural history; zoology, botany, and geology* 9(6):49–54.

2. Acosta, L.E. (2015). 'Historia de la zoología en la universidad de Córdoba: los primeros años (1872-1916)'. *Revista Facultad de Ciencias Exactas, Físicas y Naturales* 2:75–95.
3. Frenzel, J. (1892). 'The mesozoon *Salinella*'.
4. Ochsenius, H.C. (1911). 'Salpeterablagerungen in Chile'. *Zeitschrift der Deutschen Geologischen Gesellschaft* 63:35–43.
5. Frenzel, J. (1892). 'Untersuchungen über die mikroskopische Fauna Argentiniens'.
6. Ibid.
7. Rumsfeld, D. (2002). 'U.S. Department of Defense (DoD) Feb 12th news briefing', https://web.archive.org/web/20160406235718/http://archive.defense.gov/Transcripts/Transcript.aspx?TranscriptID=2636
8. Locey, K.J. and Lennon, J.T. (2016). 'Scaling laws predict global microbial diversity'. *Proceedings of the National Academy of Sciences USA* 113:5970–5.
9. Arroyo, A.S. et al. (2016). 'Hidden diversity of Acoelomorpha revealed through metabarcoding'. *Biology Letters* 12:20160674.
10. Wolf, E.T. and Toon, O.B. (2015). 'The evolution of habitable climates under the brightening Sun'. *Journal of Geophysical Research: Atmospheres* 120:5775–94.
11. Farnsworth, A. et al. (2023). 'Climate extremes likely to drive land mammal extinction during next supercontinent assembly'. *Nature Geoscience* 16:901–8.

Index

Index